CLEAN

ALSO BY JAMES HAMBLIN

If Our Bodies Could Talk:
A Guide to Operating and Maintaining a Human Body

CLEAN

THE NEW SCIENCE OF SKIN AND THE BEAUTY OF DOING LESS

JAMES HAMBLIN

RIVERHEAD BOOKS ··· NEW YORK

RIVERHEAD BOOKS
An imprint of Penguin Random House LLC
penguinrandomhouse.com

The Library of Congress has catalogued the Riverhead hardcover edition as follows:

Names: Hamblin, James, author.
Title: Clean : the new science of skin / James Hamblin.
Description: New York : Riverhead Books, 2020. |
Includes bibliographical references and index. |
Identifiers: LCCN 2020007581 (print) | LCCN 2020007582 (ebook) |
ISBN 9780525538318 (hardcover) | ISBN 9780525538332 (ebook)
Subjects: LCSH: Hygiene. | Beauty, Personal. |
Skin–Care and hygiene. | Self-care, Health.
Classification: LCC RA776 .H1825 2020 (print) |
LCC RA776 (ebook) | DDC 613–dc23
LC record available at https://lccn.loc.gov/2020007581
LC ebook record available at https://lccn.loc.gov/2020007582

First Riverhead hardcover edition: July 2020
First Riverhead trade paperback edition: July 2021
Riverhead trade paperback ISBN: 9780525538325

Printed in the United States of America
1st Printing

BOOK DESIGN BY LUCIA BERNARD

CONTENTS

CLEAN

PROLOGUE

Five years ago, I stopped showering.

At least, by most modern definitions of the word. I still get my hair wet occasionally, but I quit shampooing or conditioning, or using soap, except on my hands. I also gave up the other personal care products—exfoliants and moisturizers and deodorants—that I had always associated with being clean.

I'm not here to recommend this approach to everyone. In a lot of ways it was terrible. But it also changed my life.

I'd like to say I stopped showering for some noble, virtuous reason—like because an average American shower uses around twenty gallons of perfectly good water. That water then gets filled with petroleum-derived detergents and with soaps made from palm oil farmed on land that would otherwise still be rainforest. The body care products transported around the world in fuel-burning ships and trains contain antimicrobial preservatives and plastic microbeads that end up in our lakes and streams and make their way into our food and groundwater and back into our own bodies. Aisles upon aisles of these products are sold in pharmacies across the globe in plastic bottles that will never biodegrade, and that end up floating together like islands in the oceans. Islands that whales try, tragically, to mate with.

The last bit about the whales is not true (I hope). But the rest of these are global effects of daily bathroom habits on the scale of seven billion people that I hadn't really considered when I first stopped showering.

For me, it started simply. It wasn't even really about showering. I had just moved to New York, where everything is smaller and more expensive and more difficult. Not long before, I'd left a career practicing medicine in Los Angeles to try becoming a journalist. Against the advice of pretty much everyone, I was transitioning from a profession that promised a half-million-dollar salary into a globally imploding job market. I had moved across the country and was back at the bottom of a professional ladder, in a studio apartment, with no clear path in any direction, much less forward or upward. A mentor told me not to start climbing again unless I knew my ladder was against the right wall.

He didn't mean "stop showering," I don't think. But I saw this as a moment to take stock of everything in my life. In the process of this existential audit, I considered the possessions and habits that I might at least try going without. I cut back on caffeine and alcohol, disconnected my cable and internet, and sold my car—limiting anything that could be an overhead, recurring, mindless cost. I toyed with living in a van, because Instagram made it look so glamorous, but was discouraged adamantly by my girlfriend and everyone else in my life.

Even though I wasn't spending a lot of money on soap and shampoo, I did think about the net amount of *time* that went into using them. Behavioral economists and productivity experts will sometimes quantify the additive effects of small choices to help people break habits. For example: If you smoke a pack a day in New York, you spend almost $5,000 a year. Over the next twenty years, with expected cost increases, quitting could save you almost $200,000. If you stopped getting so much Starbucks, as I understand it, you could have a second home in Bermuda. If you spend 30 minutes per day showering and applying products, over the course of a long life—100 years, for purposes of optimism and ease of math—you will spend 18,250 hours washing. At that rate, not showering frees up more than two years of your life.

Friends and family suggested that I would have trouble enjoying the extra time because I would feel gross, unkempt. My mother worried I'd get sick from some germs I failed to clean off. Maybe I would miss the basic humanity of the routines that compel us to take time for ourselves, and that give us at least some semblance of power to present ourselves as we wish the world would see us. Or I'd miss the simple ritual of taking a nice warm shower and emerging each morning like a new person ready to face the day.

But what if none of this happened? What if I actually got fewer colds, and my skin looked better, and I found other, better routines and rituals? What if all those products in our bathrooms—shampoos to remove oils from our hair, and conditioners to replace them; soaps to remove oils from our skin, and moisturizers to replace them—were mostly effective in getting us to buy more products? How do you really know if you've never gone more than a couple days without them?

"I know what it's like to not shower," goes the most common reply from skeptics, "and it's not good." To which I say, yes. I know what it's like as a coffee-drinker to go without coffee, and it's not good. I know what it's like to go into a party where I know no one, and it's not good. I know what it's like to try to run a marathon without training, and it's not good. But I also know what it's like to slowly use less and less caffeine, and to come to feel at home in new social circles, and to build up to running twenty-six miles without yearning for the sweet embrace of death.

The more gradually a human body eases into these endeavors, the easier they are to do and even to enjoy. Changing daily cleaning habits could be thought of the same way. Over the course of months, and then years, as I gradually used less and less, I started to need less and less—or, at least, to believe I did. My skin slowly became less oily, and I got fewer patches of eczema. I didn't smell like pine trees or lavender, but I also didn't smell like the oniony body odor that I used to get when my armpits, used to being plastered with deodorant, suddenly went a day without it. As my girlfriend put it, I smelled "like a person." Initial skepticism turned to enthusiasm.

I am under no illusions that I never smelled bad. But it happened less and less regularly. And I started to become aware of patterns. Breaking out or smelling bad usually coincided with other factors: stress, sleep deprivation, generally not thriving. Out at my family's tree farm in Wisconsin or on vacation hiking in Yellowstone, when I might go for days without indoor plumbing, I was almost guaranteed to smell and look decent. In the indolence of winter days barely moving except to get to and from the office, I felt squalid and smelled accordingly. Essentially, I became more attuned to what my body was "trying to tell me." It seemed to be telling me not so much "wash me" as "go outside, move around, be social, et cetera." (My body still sometimes trails off and says "et cetera.")

It was possible for me to stop showering in large part because I was born with extra credit in the ingrained currency of acceptability in America: I'm an ambulatory white male who appears generally healthy. I'm relatively young and can afford clothes that fit and aren't tattered (or may even purposely be), and to wash and change them regularly. I'm literate and fluent in the dominant local language. All of these things, among others, mean that I move through the world absolved from expectations to look a certain way in order to be perceived as someone who belongs. Even when I am not showered or groomed, I'm unduly likely to still be seen as competent or professional or welcome in a restaurant. In other words, I barely have to do anything to be seen as clean.

The social standards that long placed value on such things are intertwined with, among other things, the history of hygiene and sanitation. Some ideas about cleanliness are nearly universal, driven by senses of disgust and revulsion that have evolutionary roots in disease avoidance and self-preservation. But others go far beyond the science of infectious disease or toxic exposure. Routines adopted to protect ourselves from disease have become enmeshed with routines that are socially determined, passed down through complex belief systems that define where we fit into the world and help us hit the right balance of belonging and uniqueness. And even our most personal decisions about caring for our bodies have long been influenced and manipulated by larger power structures.

In the course of working on this book, I also got a degree in public health and finished a residency in preventive medicine. This relatively new specialty considers itself a counterbalance to a medical culture that has come to focus too heavily on reactive solutions and narrow, temporary treatments that leave basic causes and fundamental problems unaddressed. It focuses instead on how to prevent disease before it starts, which often comes down to ensuring people have access to basic things like decent food, clean water, and communities where they can safely lead engaged, active, purposeful lives. Health means different things to different people, but it's always associated with a certain level of freedom—especially financial and temporal—that allows people to live well, and to focus on relationships and meaningful work.

That basic philosophy made me only more curious about the money and time we collectively invest in skin care—and the standards that define what's acceptable. Many of these can be traced to an industry that has, for the past two hundred years, sold us promises of health, happiness, beauty, and all manner of acceptance based on literally superficial fixes. And so I ended up on a multiyear journey through the history and science of soap, deconstructing the fortunes, products, and belief systems it has spawned, from the "soap boom" of the nineteenth century right up to the modern skin care industry. After talking to microbiologists, allergists, geneticists, ecologists, estheticians, bar-soap enthusiasts, venture capitalists, historians, Amish people, international aid workers, and a few straight-up scam artists, I came to believe that we are at the beginning of a dramatic shift in the basic conception of what it means to be *clean*.

The global market for soaps, detergents, deodorants, and hair and skin care products is now valued in the trillions of dollars. The parade of bottles and tubes and vials that line modern bathtubs and medicine cabinets surpasses the collections of yesterday's monarchs. Much is sold to us not as luxury but as necessity. The industry has grown to unprecedented heights largely on the promise of defending our bodies from the outside world.

As the scope and intensity of global cleaning practices has escalated, we've been oblivious to their effects on the trillions of microbes that live on our skin. Scientists are only now learning just how these microbes influence processes throughout our bodies. The vast majority of our skin microbes seem to be not simply harmless but important to the skin's function and, so, to the functioning of our immune systems.

The skin microbiome represents a new and important reason to reconsider much of the received wisdom about soap and skin care, and to think deliberately about the daily habits many of us undertake in pursuit of health or well-being. The skin and its microbiome are the interface between our bodies and the natural world. Our microbes are partly us and partly not. Our growing understanding of this complex, diverse ecosystem has the potential to completely change how we think about the barrier between ourselves and our environments.

This book, in the end, is an invitation to embrace the complexity of the world around us and on our skin. Even if you don't stop showering.

• • •

I wrote this book in the years before the coronavirus pandemic, which took hold just as we were going to press, so you won't find any mention of COVID-19 in the following pages. The stories and principles I share are no less relevant in this new era of pandemic awareness, as we recover from one and brace for the next. Maybe more than ever, this is an important moment to examine our daily habits, and to be deliberate about what we consume and how we relate to the natural world. I'm hopeful that a conscientious understanding of microbial life will serve us well in the years to come.

I

IMMACULATE

I walk off the elevator into a palatial, sun-soaked office looming seven stories over Bryant Park in Manhattan. It's the fall of 2018, some three years since I last washed my face. I'm here to see what the effects have been.

The herringbone wood floors are decorated with floral bouquets as tall as any human. The fireplace has a white mantel, and ethereal flute-based music wafts over me from somewhere. A bed draped in white linens awaits under a chandelier. This is the headquarters of an up-and-blazingly-coming skin care company called Peach and Lily. It's based in the Korean tradition and part of a Westernized movement commonly known as "K-beauty" that centers on maintenance of one's skin, often through a ritual of cleansing, toning, moisturizing, and sheet masking that can include ten or more steps.

The company's founder, Alicia Yoon, holds an MBA at Harvard Business School. She is also an esthetician, perhaps best known for her work in popularizing the application of snail secretions to the skin. In just two years, Yoon took Peach and Lily from a small internet boutique to a full line of original products distributed at retailers like Urban Outfitters and CVS. The company arrived on the crest of an enormous wave. In South Korea, where K-beauty is based on long-standing tradition, the industry has exploded to more than $13 billion per year. Its newfound popularity in the U.S. has helped make skin care a faster-growing segment of the beauty industry than makeup. Sales of high-end skin care grew 13 percent in 2018 alone, significantly faster than GDP.

I'm greeted at the elevator by a cheery assistant who asks me to disrobe. I explain that I've come to have a facial. She laughs and says she knows that, then hands me a robe and a questionnaire about my skin care routine and leaves me to change.

Alone in the space, I turn to the questionnaire, which looks like a form you might fill out in the waiting room of a doctor's office. There is a question about allergies and diet, along with a battery of questions about my skin: What exfoliants do I use? What moisturizers? What serums? What cleansers? How often have I been using each and in what order and combination?

This is a brief exercise for me, since I've been doing nothing. Yoon enters and welcomes me graciously, but the tone shifts when she sees the mostly blank form and learns that I have not simply forgotten to fill it out. "Oh my god," she says. "Are you safe to have a facial?"

"Yes! Of course—wait, why would I not be?" I hadn't really considered I was at risk before. Suddenly I'm worried. "I don't know—I mean, you tell me."

"It should probably be fine, I've just never done this on anyone . . . like this before," she trails off, either sad or disappointed or maybe both.

I lie down and she puts a bright light over my face. She touches my cheek lightly with her fingertip, then a little more firmly. Hesitantly, she says, "Have you ever felt your face?"

Funny she should ask.

I have made a point of almost never touching my face, ever since I was a teenager with "bad skin" who was under the now-obsolete impression that acne is caused by not cleaning well or aggressively enough. There were times the acne would extend to my eyelid as a stye that would nearly swell my eye shut. Social interaction became impossible because the appearance of my eye sucked all the attention out of any conversation. Even after my skin cleared up in college, I held on to the habit of keeping my hands—and the bacteria and viruses they carried—away from my face.

Not wanting to get into this long backstory, I tell Yoon that I touch my face "the normal amount," and she gets to work.

Yoon is no stranger to "problem" skin. She spent much of her life battling severe eczema, at times scratching her inflamed skin until there was no skin left to scratch. "I grew up trying everything under the sun, even bleach baths," she tells me, referring to the dubious practice intended to kill any and all microbes on one's skin.

But when she attended esthetician school in Korea, she began to experiment with new skin care rituals to calm the inflammation. Finding a routine of gentle, moisturizing products was part of what she has called her "skin-care breakthrough moment," and the approach she now shares with her clients.

Yoon applies Peach and Lily's Glass Skin Refining Serum (the bottle promises "translucent + luminous" skin as well as "peptides") and Pure Beam Luxe Oil ("replenish + rebalance" with jojoba oil) to my face, as well as a Super Reboot Resurfacing Mask containing blue agave, and a Matcha Pudding Antioxidant Cream.

She recommends to me the Original Glow Sheet Mask for home use, as it contains hyaluronic acid.

Hyaluronic acid binds water, so it adds volume in the epidermis. Babies have a lot of hyaluronic acid, which is partly why their skin is smooth and firm and plump. Whether putting it *on* your skin is the same as having it *in* your skin is an open question. It does no good to cover your car in gasoline, or to fill your house with roofing shingles. According to dermatologists I've spoken with, some forms of the acid can *sometimes* penetrate into the skin, but only the ones that have smaller molecular weights. The Peach and Lily mask doesn't specify whether it's the type that penetrates the skin, but it does purport an "anti-aging" effect.

It is unlikely a coincidence that the skin care boom is happening at a time when people are losing trust in science and medicine, often for good reason. Dermatologists, like many other doctors, are typically scarce and expensive. And many people feel the profession has failed them. If it can ever be said that skin is "good" or "bad"—meaning uncomfortable, dry, irritated, itchy, painful, or otherwise causing us distress—then our collective skin is getting worse. Rates of the inflammatory skin condition known as atopic dermatitis, or eczema, are increasing rapidly. According to the World Health Organization, the prevalence of psoriasis more than doubled between 1979 and 2008. Acne continues to afflict people during prime years of social development, and research suggests it is also increasing in adults, especially among women.

The causes of these trends are complex, going far beyond the skin itself. For example, a 2018 review of studies in *Clinical, Cosmetic and Investigational Dermatology* suggested that one reason for the increase in acne in women is the hormonal imbalances associated with "metabolic syndrome"—the term for the constellation of diabetes, cardiovascular disease, and obesity. High insulin levels can cause the body to convert estrogen to testosterone, which signals growth factors in the skin that lead to more oil being secreted, changing bacterial populations and fueling a cycle of inflammation whose culmination is a pimple.

With elaborate processes like these underlying the appearance of skin, it's no wonder that topical treatments alone, for acne and other common skin conditions, often work incompletely or unreliably. Systemic treatments are rarely less fickle. Oral contraceptives are sometimes prescribed in an attempt to even out an alleged hormonal imbalance. People tolerate these recalibrations very differently, and results vary from life-changing to useless—while coming with the side effects of changing one's entire body chemistry solely to address the skin. Antibiotics don't reliably help either, and powerful drugs like Accutane carry the potential to cause birth defects and, many users report, intense depression. In psoriasis and eczema, a person may cycle on and off steroid treatments for much of a life time, never finding a definitive cure or even knowing when or why a flare will occur. The overall effect of such trial and error can leave patients believing they might as well take matters into their own hands.

Desire for control and certainty also leaves people wanting preventive approaches, of the sort that the medical system has not traditionally taken seriously. Yoon is seeing growing demand for products that promise to "nourish" or "protect" the skin. This is also partly due to growing concerns about air pollution, she hears from consumers, and about increasingly intense ultraviolet radiation from the sun as greenhouse gases dissolve the ozone layer. As the Earth loses this protective shell, people are compelled to apply their own.

Each of the products Yoon applies to my face promises some amalgamation of attractiveness, protection, and maintenance—blurring the line between cosmetic enhancement and essential defense against *toxins* and other ambient dangers. She likens some of the rituals to providing nutrition for my face, saying they will help "make sure your skin has the vitamins and minerals and fatty acids to thrive." I begin to feel negligent.

Cosmetics are not food, legally. They are also distinct, in a regulatory sense, from drugs in that they can't claim to treat or prevent specific diseases. But sellers *can* market these products with claims about improving and maintaining health—without all the bureaucratic burden of getting a drug approved to be sold on the market. Yoon is part of a new generation of entrepreneurs who exist in a place that is not quite health or beauty but a mix of both. New skin care products promise to obviate the need for makeup and medicines by making the skin look "naturally" good. These products are not simply offering a recreational way of temporarily altering our appearance but are getting much closer to what would usually be considered drugs—implying that they are preventing or fixing a problem in the functioning of the skin.

The new industry bypasses traditional gatekeepers, as products can be aggressively marketed right in our Instagram feeds. YouTube influencers build personal brands based on anti-establishment solutions to skin "problems," speaking with the kind of persuasive personalities that medical school seems designed to kill off. Here, everyone is an expert. No mountain of studies is likely to override what worked or didn't work for you.

If you've ever been unhappy with your skin, the appeal of these sorts of promises is probably clear. When the antibiotics recommended by my dermatologist didn't help with my teenage acne, my innovative dentist father even suggested that since trying yet more oral antibiotics could have undesirable side effects, I might apply them topically. So I took tetracycline capsules and broke them open, mixed them in water, and rubbed that all over my face. Then, instead of simply redness and lumpiness, I had a yellowish hue. People asked if I had been using some kind of tanning spray, as some midwestern teenagers did at the time to simulate the appearance of having gone outside. I laughed and told them that was ridiculous. In fact I had tried that, along with most anything else, to attempt to even out the very weird palette of my face. But you know what happens when you add orange to red and yellow? You get a weirder orange. Even less natural-looking, even more disconcerting.

Lying on the crisp linens at Peach and Lily, above the noise and impersonality of the city streets below, I'm not thinking about marketing or my teenage angst—or much at all. If you have never had your face massaged, let me assure you it is wonderful. More than just the physical sensation of the massage and the application of products, it's an act that takes you instantly out of whatever stressful things are happening in your life, into a feeling of being temporary royalty. Another human is taking the time and effort to *rub your face*, simply to make you feel and look good.

When the renovation is complete, Yoon loads up a bag full of samples for me to experiment with at home. She can't give me any of the Glass Skin Refining Serum because it's out of stock everywhere and even she doesn't have enough. It sold out instantly upon being announced.

"You need to take care of your face," she tells me, urging me to at least use a cleanser. I laugh; she doesn't. I flush. As I get into the elevator, she says again, firmly: "More needs to be done."

When I emerge onto the street post-facial, from what was apparently a cocoon of dead skin and oil on my face (who knew?), I experience the world differently. I step out into the sunlight and—this might be hard to believe if you've never gone years without cleaning your face and then had a wildly fancy facial—I can feel the world with my face in a way I didn't know was possible. My skin—I touched it—is definitely softer. And though it may only be in my head, I instantly feel like I am being seen differently. Maybe because of some kind of newly confident spring in my step, or because I actually look more attractive. Maybe I simply look like a person with the means to put matcha on my face.

In any case, I feel changed. Sometimes that's a perfectly sufficient thing to want. Not necessarily to be better, just different. I'm reminded of how easy it can be to get used to the way the world treats us, and come to conceive of our place in it accordingly. Once we do, it's easy to take notice of only the outlier experiences—when people are nicer or meaner to us than we've come to think we deserve. This effect is also drawn out in other moments of dramatic physical change like getting really dressed up or getting a radically different haircut. The very real ways that appearances inform the way people treat other people are uncomfortably palpable in those moments.

The other change I feel will be more lasting. Up until this moment, I had been fine without facials for my entire life. If they had ever even crossed my mind, I probably would have dismissed them as self-indulgent vanity and, if I'm honest, as a child of Indiana, not something men *do*. At the very least, facials weren't something I was interested in spending my time and money on. But seeing how something as simple as someone rubbing things into my face could change the way I move through the entire day, the sense of frivolity vanishes. I see how, like so many things that feel extravagant the first time, these serums and oils and masks could lose their initial feeling of luxury and start to feel routine, even necessary.

Many of the cleaning habits we now take for granted started relatively recently. Over the course of just a few centuries, social and personal standards for hygiene and cleanliness in much of the world have expanded from an occasional jump in the river to an essential daily shower or bath. Now to even speak of not showering is, as it's been put to me, "not really dinner conversation."

Bouncing between worlds of minimalism and maximalism made me curious about an ideal balance. I didn't want to start another expensive habit. (And don't the snails need their mucus?) But I also didn't want to be missing out on something that clearly brings people a lot of joy, and can change the course of day-to-day interactions in meaningful ways. What should I be doing to take care of my skin? How much of what people do is about enjoyment—or at least not disgusting other people or seeming negligent or oblivious—and how much might actually improve my own health and well-being?

In any case, it was going to be tough to go back to doing nothing.

• • •

I've never experienced such a balanced mix of love, disgust, curiosity, and vitriol as I did when I wrote a short article for *The Atlantic* in 2016 about how I had stopped showering. Readers wrote to me by the hundreds to express feelings across the emotional spectrum: to tell me they'd figured out what I figured out long ago, to tell me I was crazy, and to get a sense of whether what they were doing, hygiene-wise, was medically okay.

Some readers hated that a doctor could be so irresponsible as to imply that hygiene didn't matter, as they read it, given the ongoing outbreaks of cholera and annual deaths in the hundreds of thousands from influenza. Others were angry that I had not made it clear that not showering was my privilege as a white male in a wealthy country.

Others thought it was totally obvious. A woman from Germany named Patricia wrote, "I couldn't agree with you more!" Hers was a compulsory detox. She went to the hospital with excruciating back pain on Easter Sunday in 2007 and was told she had had a stroke. "With 1.5 hands, showering is work," she wrote. "I did ask friends and neighbors to 'pls tell me if it smells here!!'" But otherwise, "All was and is fine. Apart from the odd 'cat wash,' showering is reduced to once a month or so." Her feet stopped smelling, and she noticed that her skin and hair seemed to produce less oil over time, allowing her to go longer and longer between washes.

An eighty-nine-year-old woman named Claire who wrote from Ontario said that she and her husband (who died at age ninety-six) never bathed. She saw it as part of a general approach to health, and she attached a photo as evidence that she looks younger than her age. She wore a white visor and shorts and was waving at the camera: "Because of my extraordinary level of exceptional health, maybe because I exercise and eat VERY selectively, I astonish all who meet me," she wrote. "I shoveled the drive twice yesterday and didn't even feel tired."

I wrote back to ask how she came to the idea of not showering. "Well, why were we washing so much?" she asked. "Didn't we have wonderful skin which flakes off all the time and cleans itself, and doesn't soap take the oil out of our skin?" She saw all of this as part of a basic life philosophy that has become popular of late. She suggested that I "eat like a cave man."

Yes, Claire was a Paleo diet originalist. Her "cave man" idea came up often in the responses I received: essentially that modern life is the cause of chronic disease, and if we followed a "Paleo diet" and ate mostly beef and butter, rejecting technology that resulted from the dawn of agriculture, we'd be fine. Though, of course, during the Paleolithic era, human life spans were much shorter than they are today. And there were no cows.

Paleolithic life was not without its perks. Humans of the time lived in such sparsely populated areas, in such small communities and caves, that they could use the waterways as their toilets without issue. Many could hunt and gather without depleting resources. In the process they were exposed to the elements—to sunlight and heat and cold, to soil and animals, and to other people who were not in any modern sense "clean."

This way of living was possible until very recently, in the scope of human history. Even as recently as 1600, the entire city of London had around 200,000 people. By World War II it had increased to 8.6 million. There are that many people today in New York City, too. The indoor area in Manhattan is now nearly three times as large as the island itself.

Each of these vertically oriented conglomerations of humans is a radical living experiment in concentrating resources and people. Global average life expectancy is now around seventy-two years. Each of us is expected to regularly use energy and transportation and the products of industrial agriculture, which involve killing trees or burning fossil fuels that fill our skies with smog and particulate matter. This makes its way into the deepest parts of our lungs, and is a leading cause of cancers and heart disease. The World Health Organization estimates that inhaling pollution is a cause of seven million annual deaths.

If there was relatively little chronic disease in the Paleolithic era, that was in part because so many people died of infections and injuries. Over the past two centuries, in most of the world, the chances of dying of an infectious disease have plummeted. Meanwhile, odds of dying of a chronic disease are vastly greater than they used to be. Globally, the number of deaths due to chronic disease is fast approaching three out of four.

For all the benefits of modern medicine and technology, the new systems of living are implicated in health issues that were once much less common. Autoimmune diseases, diabetes, and cardiovascular disease are all on the rise, at least in part because so many people now live longer than in generations past. But these chronic diseases are also afflicting younger people at high rates, too, suggesting they are also related to our lifestyles and environments.

In recent years, much attention has been paid to the roles of the food system and sedentary lifestyles in chronic diseases. The importance of other environmental factors is less widely acknowledged. Among them is the fact that in much of the world, people now live most of their lives indoors, in climate-controlled environments where there is no dirt, and few plants and animals. Windows remain shut except on the most perfect of days. In these and many other ways, most people are removed from many exposures that were once common.

This distance is sometimes necessary. In 2019, as smog enveloped Delhi, millions of people were advised to stay indoors and avoid physical activity for days at a time. Such pollution events—as well as outbreaks of infectious diseases that require distancing—will likely occur more and more frequently, and in more and more places.

Whether by necessity or preference, an increasingly isolated, indoor way of living seems to have played a role in altering the functioning of our immune systems—and our primary immune organ, the skin—in ways we are just beginning to understand. For most of human history, a steady barrage of exposures to microbes would train our immune systems to know when and how to react. Today, an evolutionarily novel set of environmental inputs has left many of our immune systems confused, unable to distinguish what should and shouldn't cause our skin to flare up. This is not unrelated to the fact that many of us are taught that it is healthy and even necessary to wash ourselves elaborately, daily, sometimes multiple times. Even in places where the risk of infectious diseases is low, we are taught to maintain undue focus on preventing them. We are expected to carry no visible dirt or mud or dust lest we be considered derelict, lazy, unattractive, unsophisticated, impolite, unprofessional. In a word: unclean.

• • •

It's usually in October, when the Canadian air starts drying out, that the men flock to Sandy Skotnicki's office. The men are itching.

Skotnicki has a comprehensive perspective on skin. She trained as a microbiologist before becoming a professor of dermatology and occupational and environmental health at the University of Toronto. She has been practicing dermatology for twenty years, always with an eye to the effects of our environments—including microbes—on our skin health.

"I say to them, 'How do you shower?'" she tells me. The men want to blame the changing seasons, as if human skin were only meant to function normally in summer. But she makes them talk about their cleaning routines. "They take the squeegee thing and wash their whole body with some sort of 'men's body wash.' They're showering twice a day because they're working out. As soon as I get them to stop doing that and just tell them to wash their bits, they're totally fine."

I ask about "bits."

"Bits would be underarms, groin, feet," she says. "So, when you're in the shower, or when you're in the bath, do you need to wash here?" She points to her forearm. "No."

Her distress is palpable as she explains how much of her career as a doctor involves pleading with men not to lather their bodies with shower gel. She tells them that moisturizing in many cases becomes necessary only because people are too far into the over-washing cycle.

Even the effect of water alone, applied to the skin, is not zero. Water, especially when hot, slowly strips away the oils secreted by our glands to keep moisture in. Anything that leaves the skin drier and more porous heightens the potential for reactions to irritants and allergens.

Skotnicki believes this is part of how over-washing harms the skin, making people who have a genetic predisposition to eczema more likely to develop flares. While eczema itself can be debilitating, it often does not travel alone. The condition seems to be part of a constellation of conditions that result from immune-system misfires. Roughly half of kids who have severe eczema will go on to develop allergic rhinitis or asthma, part of a cascade of overreactions of the immune system known as an "atopic march."

The concept was first described by allergists at the University of Pennsylvania and University of Chicago in 2003, when they noticed these patterns in children. Since then the associations have continued to bear out. Studies have even implicated the recent rise of peanut allergies. In 2010, allergists at King's College London said they were "shocked to find out" that infants with asthma were more likely than their peers to have peanut allergies. By 2019, the director of the National Institute of Allergy and Infectious Diseases, Anthony Fauci, advised parents that "early intervention to protect the skin may be one key to preventing food allergy."

The idea of skin care to prevent food allergies is still not fully understood, but recent recommendations now suggest that exposing young children to peanuts—as opposed to sheltering them—may decrease their likelihood of developing a severe peanut allergy. As with the vaccination shots that doctors give to train the immune system to fight various infectious diseases, exposure to small amounts of peanuts is believed to train the immune system through exposure. Yet still essentially the opposite is done with immune-related conditions of the skin. Many treatment approaches involve immune-suppressing drugs, antibiotics, and, of course, aggressive cleaning and moisturizing regimens.

Eczema is so common that it's often dismissed as a mild annoyance—which many cases are. But the condition can also render a person acutely miserable. It can affect a person's sleep (most itching happens at night) and livelihood, their basic ability to do anything without scratching. The condition seems to bring together everything that can go wrong with the skin: disrupted barrier function, microbial imbalances, and immune cell amplification. Perturbing the skin barrier by washing or scratching can change the microbial population. That can rev up the immune system, which tells the skin cells to proliferate rapidly and fill with inflammatory proteins. All of this is a self-perpetuating cycle of inflammation, itch, barrier breakdowns, and microbial imbalances. “And so what if,” Skotnicki speculates, “as a society, you actually created eczema by over-washing?”

The two have at least increased in tandem, and there is evidence that their rise is not unrelated. Instead of increasing our exposures, allergies and hypersensitivities have only led us to do *more* cleaning and sterilizing of our environments. When patients come to see Skotnicki, often with rashes that last weeks or months, their instinct is to do yet more scrubbing and soaping. They come to her hoping for another product—something to undo or perhaps counterbalance the current products. They want something “mild and natural.” They want something that’s, well, as close as possible to nothing at all.

It’s difficult for doctors to prescribe nothing. Patients often want a treatment—if not a prescription, at least something that can be done in a regimented way. So Skotnicki has found a way to turn nothing into something. She advocates a total product “diet” or “cleanse”—as in, stopping everything. (Or, as much as possible.) This approach is increasingly supported by dermatologists as a conceptual reset, even if no particular product was causing a clear problem. It can be psychologically valuable to see how little we actually need, and to slowly reintroduce only the things we really want.

Skin is, after all, extremely resilient. We can try to control or coat it with topical products, but it is ultimately a force of nature reacting to the constant signals coming from underneath and outside of it, as it evolved over millions of years to do. It is trying to maintain equilibrium.

• • •

Skin is the human body's largest organ. We each have enough skin that if you lay it out, it would cover around twenty square feet. It can move in any direction and stretch and sense tiny changes in temperature and pressure and moisture. Skin contains the tips of nerve fibers that can send signals to our brain to create excruciating pain and ecstatic pleasure. Skin communicates to the world when we are sick or tired or anxious or aroused. Skin can be torn wide open and heal itself back together in a few days' time. Skin can keep us from fatally overheating by drenching itself in liquid that causes heat to radiate more quickly into the adjacent air. Skin is no less vital than our heart or spine or brain. Without it the fluids that compose us evaporate, and the outside world pours into us and infects us and we quickly die.

So skin care is extremely important. But effective skin care goes far beyond applying things to its surface.

The textbook understanding of how skin works—and what I was taught in medical school—is that skin consists of three anatomical layers. The bottom one is mostly fat and connective tissue. The other two are more interesting. The outermost is the epidermis. It's about a millimeter thick, like a sheet of paper, but an amazing amount happens in that millimeter. The primary cells of the epidermis are called keratinocytes. These produce the keratin protein that makes up most of our skin and the entirety of our fingernails and hair. Intermixed with these proteins is a collage of immune cells and tiny nerve fibers, as well as cells that produce the melanin that gives all skin its color. All of these cells are extremely sensitive to our environment, and they react and change accordingly.

The epidermis is constantly regenerating itself, as almost no other part of the body can. The millimeter-thick layer is itself divided into strata representing cells of different ages. The basal layer contains stem cells, which continually divide, producing new cells. That process happens more readily in youth. But throughout life, the skin is always generating new cells that push older ones up toward the surface. By the time they get there, they are mostly dead, flat, dehydrated, and stuck together such that they are visible to the naked eye. The goal of exfoliating products is to take off this outer foliage and expose the newer cells to the world, though the cells will also shed naturally. The entire cycle takes about a month and serves to continually rebuild the surface of the skin.

Below the epidermis is the dermis, a layer mainly composed of two proteins: collagen and elastin. Woven together, they give skin its elasticity and strength. Leather, for example, is pure dermis. The inimitable mix of pliability and durability is why, despite the enormous cost and ethical concerns of hunting down animals and taking their skin, humans have insisted on using leather to protect ourselves and survive the elements since before the advent of tools.

Coursing through the epidermis and the dermis are networks of nerves that can detect even the slightest changes in our environment, discerning the weight of a mosquito or the difference between a 68-degree and a 72-degree office. This network is crosshatched with microscopic blood vessels that expand to cool the body during exercise and stress, and that cause us to flush and make our emotions manifest to the world.

There are also clusters of relatively enormous structures called follicles. These create our hair, which allowed pre-human species to move into cold climates, and for which there is now an enormous market for removal and shortening, contouring and coloring, according to norms that signify where people fit into a social hierarchy, and where they want to belong.

The skin also contains three types of glands that secrete oils and other compounds. The basic sweat glands (known as eccrine glands) secrete water to cool the body. Sebaceous glands secrete oily sebum that lubricates the skin—so that we don't dry out and crack open, compromising the barrier and allowing microbes in, causing death.

Less obviously explicable are the apocrine sweat glands that develop during puberty, especially around the armpits and groin. They add oily secretions of their own, which to many people seems excessive—even cruel. These are the glands that we attempt to block with antiperspirants, and whose existence many of us spend much of our lives struggling to counteract. We are now learning that these glands are involved in sustaining another important part of our skin, which could amount to a fourth layer: the trillions of microbes that live in and on us. The airborne chemicals that account for our bodily odors are a product of bacteria on our skin, especially in the armpits and groin, feeding on our oils.

These microbial populations are influenced by the amounts and types of oils we exude, as well as other compounds like sodium, urea, and lactate that pour out of us when we sweat. Sweat has also recently been found to contain peptides with antimicrobial properties, like dermicidin, cathelicidin, and lactoferrin. These compounds seem to have some part in maintaining and restoring microbial balances. If you ever feel self-conscious about sweating, you might explain to those around you that your body is simply partaking in an elaborate and mysterious biochemical ballet.

The fact that we carry around some microbes has long been

known—for as long as scientists have been able to grow cultures of bacteria, they have known that swabbing human skin is a reliable way to start a prodigious microbial garden. But only over the course of the past decade has new DNA sequencing technology begun to reveal the scale and diversity of microbial life. The microbes on our skin, combined with those in our digestive tracts, account for several pounds of our body weight. There are more microbial cells in and on each of us than human cells.

While we have long thought about our skin as a barrier to separate us from the outside world, growing knowledge about the microbiome suggests that skin is instead a dynamic interface with our environment. These microbial ecosystems really amount to extensions of ourselves. Like the microbes that fill our guts, the microbes on our skin rarely cause disease. If anything, they may help protect us from disease. And everything we do—and don't do—to our skin has some effect on these populations.

When we clean ourselves, we at least temporarily alter the microscopic populations—either by removing them or by altering the resources available to them. Even if we do not use cleaning products that specifically say they are "antimicrobial," any chemistry applied to the skin will have some effect on the environment in which the microbes grow. Soaps and astringents meant to make us drier and less oily also remove the sebum on which microbes feed.

Because scientists and doctors didn't have the technology to fully understand the number or importance of these microbes until recently, very little is known about what exactly they're doing there. But as this new research elucidates the interplay of microbes and skin, it is challenging long-held beliefs about what is good and bad.

• • •

There may be no more memorable case of skin microbes changing our understanding of ourselves than the mites that live on our faces.

In 2014, a group of researchers swabbed the faces of 400 volunteers in North Carolina and discovered microscopic mites called *Demodex* living on their skin. Usually burrowed in our pores, the half-millimeter demon arachnids are colorless and have four pairs of legs that are all on the anterior third of their bodies, the rest of which drag behind them. Somehow, as a Swiss dermatology journal described the mites' anatomy (possibly addressing some concern about what mites might be doing on our faces), "an anus is lacking." Anus or not, the initial response from me and so many others was: *Good lord, get these off me instantly.* The more staid science journalists ran headlines like the one on NPR's site: "Hey, You've Got Mites Living on Your Face. And I Do, Too."

Of all our microbes, the mites are the only ones large enough to see with a magnifying glass (that we know of). Below them in size are fungi, which are scarce in living people due to our body temperatures. Then come bacteria, archaea, protozoa, and then much smaller viruses. So a real mystery of the mites is why they aren't better known. These mites were actually discovered long ago—in 1841, when a German anatomist first found them on some corpses, and then occasionally on live humans. Though he documented the finding and wrote that it could be important, the tiny mites were largely forgotten.

So why did the North Carolina mite hunters just now find that *Demodex* are all over all of us?

The effort was made entirely possible by the new DNA sequencing technology that discovered the rest of the microbiome. The actual mites are hard to find, as they're often burrowed deep in our pores. If you look for evidence of their DNA on our skin, though, we all have it. This technology is the reason we're only just learning about our little comrades—among many, many others.

Upsetting as it tends to be for people to learn about their mites, it would hypothetically be worse *not* to have them. When something is a feature of 100 percent of people, this is as close as we will ever get to a proper definition of "normal." They must be there for some purpose. Right?

Michelle Trautwein, an endowed chair of dipterology (the study of flies) at the California Academy of Sciences and a coauthor of the study, sees a sort of existential beauty in the mites: "They're a universal part of being human." Solving the mystery of *why* we have them is the reason that an insect biologist like Trautwein is currently working with dermatologists and ecologists, elucidating broader truths about ourselves. For one: we humans are not biologically self-sufficient organisms, but covered in and surrounded by other organisms, on which we rely.

Trautwein says the mites may actually feed off our dead skin cells—making our skin microbes the most "natural" exfoliants of all. This would mean they could decrease the amount of dust in our homes, which is partly made up of skin cells. And yet if you saw a product at the drugstore or on Instagram that promised to rid you of face mites, that would be an enticing pitch.

Though we all have mites on our faces, there is evidence that an abnormal proliferation—or abnormal reaction to said proliferation—can result in skin diseases. A recent analysis of forty-eight studies found an association between the density of mites and rosacea. Like so many microbe-related diseases, this relationship seems to be all about ratios and context—not simply about invasion by a "bad" organism. While *Demodex* are normally benign—or possibly doing something beneficial—they can become pathogenic (disease-causing) when their context changes. It's sort of like how people are rarely born with an inclination to harm other people, but many will not hesitate to kill when dropped into an active combat zone and ordered to open fire.

So these mites and the trillions of other tiny creatures that make up the skin microbiome are upending the traditional conception of "germ theory"—the simple idea that we must fight off microbes in order to avoid disease. This is being supplanted in popular understanding by a much more interesting picture. Most microbes are not just harmless but supportive to us, even vital. Self and other is less of a dichotomy than a continuum.

Though babies develop in a sterile environment—the uterus is without microorganisms—a newborn emerges as a sort of squalling bacterial sponge, and begins picking up microbes that contribute to its health and ability to survive immediately upon its passage through the birth canal. The skin is populated then by the mother's bacteria, some of which will remain for life in the pores, mediating interactions with all the other microbes the person encounters.

From that point on, the health of skin is all about context. The microbes are influenced by the outside world above them and the skin below, and the skin is influenced by the microbes above it and the bodily functions beneath it.

Research into the microbiome seems poised to overturn even our most basic assumptions about how to take care of our skin—and its implications are far from superficial.

Take, for example, a recent study led by University of California,

San Diego, dermatologist Richard Gallo. His team covered a group of mice with the bacterium *Staphylococcus epidermidis*, which normally occurs on most human skin. They cleaned other mice so they had no such bacteria.

Then they gave both groups of mice suntans. Those with the bacteria got fewer skin cancers. The reason, Gallo theorized, is that this skin bacterium produces a compound called 6-N-hydroxyaminopurine that seems to target tumor cells and prevent their DNA from replicating.

This is an early study—in microbes on mice, not microbes on humans. (It's unethical to expose humans to ultraviolet light and see if they get cancer.) But more like it now seem to be coming out weekly. Together they at least raise the question of whether we should be cleaning bacteria off our skin as aggressively and indiscriminately as many of us were taught was necessary.

To answer this requires exploring how we arrived at modern notions of what it means to be clean.

II

PURIFY

Val Curtis shows strangers images of rotten food, worms, bodily fluids, and other things of that nature. Then she records their reactions.

This is work for her. In the course of becoming the world's leading "disgustologist," Curtis started doing research as a professor at the London School of Hygiene and Tropical Medicine to understand why people care—often in deep, visceral, passionate ways—about being clean.

Curtis's research finds that people's reactions to images like these are extremely similar, almost universal—across countries and ages and genders and every other recorded variable. She distills the common response to the "filthy, sticky, oozing, teeming matter" in her studies as "a powerful feeling of disgust."

But what is behind this feeling? Curtis deployed a consumer-research technique called "laddering," which is used to help people articulate their deeper motives. The technique involves simply asking, in the manner of three-year-olds everywhere: Why, why, why? When you ask a person in a restaurant why they ordered a particular salad, for example, they might say, "It sounded good." But if you continue to ask "Why?" eventually you will get into all the complex relationships we have with food and our own mortality and our control over it. Laddering is good for first dates, as well as research. In the case of Curtis's inquiries, the answers eventually circled back to the same word: "disgust."

"Dirt is just *disgusting*. Muck is *disgusting*. Spoiled food is *disgusting*," she tells me. "I couldn't get any further than that."

So she set out to see what these things have in common.

Curtis transformed her office into a compendium of books and articles about the objects of her research—a "huge, motley collection of things that people around the world found disgusting," she says. As she started looking for patterns, "it all came back to disease."

A fallen hair, for example, can transmit ringworm. This may be why a single errant strand on a dinner plate can cause a person to condemn an entire restaurant, and to never set foot on the premises again, and to place a curse on the family of the chef.

Vomit, Curtis says, another common object of disgust in her research, "can transmit about thirty different sorts of infections."

It's not the suffering that disgusts us, it seems. If someone is dying of cancer or having a heart attack, we have no aversion to rushing to their side. Instead, Curtis suggests, seeing blood or vomit or feces or leaking wounds—all carriers of pathological microbes—triggers an instinctive aversion to protect ourselves from infectious diseases.

"Probably the riskiest thing you can do in your daily life is to come into contact with someone else," Curtis explains, "because other people are what carry the bugs that are going to make you sick."

In that way, disgust is a useful mechanism. We guard ourselves from other people's diseases by being disgusted by their behavior or appearance. It's also why we can be disgusted by ourselves, or why we might feel shame and embarrassment over how we look: the risk of social isolation and exclusion from our community motivates us to make ourselves not-disgusting to others. We evolved to care about appearances.

"If you want to be my friend, you have to be able to look in my eyes and hear me, to shake my hand, to share bodily fluids to a certain extent—because we'll breathe on each other," Curtis says. "If I were dirty and unkempt and had parasites all over my skin and lots of bodily lesions and I smelled bad, you're going to be disgusted by me. As a result I don't get the benefit of being part of your society.

"This is dangerous," she adds. "We're a collaborative species and we need each other in order to survive."

Life is a constant tension between the need to be close to other people and the need to protect ourselves from other people.

The motivation to perform "hygiene behaviors," as evolutionary biologists refer to cleaning, is seen throughout the animal kingdom. Caribbean spiny lobsters have been shown to avoid peers with viral infections. Ants groom themselves to remove disease-causing fungi, and dispose of the corpses of their fallen brethren. Bees remove their diseased friends from the hive and leave them to die. This may seem cruel, but they don't have elaborate, modern health care systems that allow them to care for their sick.

It appears that all vertebrates practice hygiene. Curtis describes bullfrog tadpoles avoiding others who have *Candida* fungal infections; whitefish can sense and avoid the parasite *Pseudomonas fluorescens*; bats groom to remove parasites, as do most other mammals, and birds. The aphorism about not shitting in your own nest is not purely a metaphor. Birds follow this advice, even on temptingly frigid days. (Instead, they defecate while flying over human heads.) Other animals have designated "latrine sites"—namely raccoons, badgers, lemurs, and others who seem to have life figured out. Chimpanzees sometimes perform what appears to be penile hygiene after mating—which is at least a nice idea, if not clearly effective at avoiding any known sexually transmitted infection.

Across the natural world, disease-avoidance behavior is as universal as love—actually more so. Even brainless nematodes that eschew love in all forms have proven able to sense and evade disease-causing bacteria. The dispassionate process of evolution was their teacher, and the genes of animals who failed to defend themselves against disease were eliminated. Those with good hygiene survived and multiplied and feasted on their fallen breatheren. No, they buried them.

In academic usage, "hygiene" technically means disease-avoidance behaviors. For humans, this means things like washing our hands, covering our coughs and sneezes and open wounds, and disposing of our feces in an orderly manner. Primal instincts to avoid disease also create and feed into existing discriminatory practices, though. Curtis explains that even in modern times, atypical appearances of people—be they limps or asymmetries or sizes that are a few standard deviations above or below the mean—may still trigger evolutionary aversions related to contamination and self-protection.

In the past, people who were swollen might be carrying diseases like filariasis, for example—the worm infection spread by mosquitoes that causes body parts to swell and skin to thicken—and so might pose a threat. Some such instincts may still manifest as aversions that accumulate to define what's considered normal. Deviating too far from the normal range—in terms of appearance or smell or sound that others perceive—continues to have social consequences, even if most of those evolutionary cues are now irrelevant.

Even while infectious diseases have been eclipsed by chronic disease as the leading causes of death, our brains still disproportionately fear infections. As disgust for ourselves and others gets mixed up by cues that actually have nothing to do with disease, it is easy to lose sight of what's really a threat. The impulse to look not-disgusting may be a foundation for what drives some modern skin care practices, though these tend to go far beyond making sure we're not covered in blood or feces.

The vast majority of what people in wealthy countries now consider hygiene is actually, Curtis explains, the pursuit of an abstract idea of *cleanliness*. Unlike hygiene, being "clean" is not just about disease avoidance.

"What most people buy hygiene products for is not the rational health benefit," she says. "It's making you look good. If you can get rid of your acne and eczema and wrinkles and smell nice—that's what people are after."

The reasons that people care about looking and smelling good are, of course, complex. Cultural standards and expectations drive behaviors that many people would happily abandon if they felt they could. Professional and social positions determine how much choice we feel we have with regard to fitting into certain aesthetic norms. Grooming has been shown to have an effect on earning power, for women in particular, as well as on overall personal body image. There is also pleasure to be taken in the ritual, in taking a few moments out of the day to care for ourselves.

Beauty can also be an end in itself. I'm advised by multiple trusted literary people that invoking Charles Darwin in a book of this sort is a cringeworthy cliché. Instead we will speak only of a nebulous nineteenth-century figure who loved finches. Even though this fellow was a chaste and homely figure in an era that prized sexual repression, his aesthetic view of sexual selection was radical. Essentially, he argued that beauty is an evolutionary trait because it gives pleasure to individuals—and that pleasure is an end unto itself. It is not something that exists just to attract mates for purposes of procreation. We animals like things that make us feel good, even if they are detrimental to long-term survival, and that includes mating with beautiful animals who are bad for us and don't stand to be good providers or even stay alive.

The much stuffier Alfred Russel Wallace ("codiscoverer" of evolution) was an antagonist of this theory, and his own argument that beauty must be the result of *adaptation*—that it exists to further the survival of the species—came to dominate science textbooks for generations. Many adaptationist theories of natural selection were based almost entirely on how men could procure women as mating partners, and how women could make themselves desirable to men. The theories were devoid of even the possibility that women are autonomous entities with a capacity for and interest in sexual pleasure.

Yale evolutionary ornithologist Richard Prum has dedicated his career to reviving the initial, buried theory of beauty as an intrinsic good. In what he calls the "beauty happens hypothesis," Prum posits that beauty began randomly, like any evolutionary process. One color or song or body size or shape or texture came to be valued, not for any reason other than that the trait brought pleasure. This preference spread socially and genetically. Instead of the explanation that males tend to be larger and more aggressive than females because they evolved to need to physically dominate other males for mating opportunities, what if females *prefer* large, powerful males? Simply because these traits are beautiful?

Prum cites the orgasm as an example of how the ability to bestow pleasure can also be advantageous to survival: The females who most *enjoy* mating are most likely to procreate. The males who can best confer that pleasure are most likely to get that opportunity. Though Prum's papers were initially rejected by peer-reviewed journals, the scientific community is finally coming around to the idea that beauty exists as a valuable entity in and of itself—even if it does not necessarily mean a person is more fit, or healthy, or reproductively viable.

Though it's taken biologists awhile to come around to this idea, the author Toni Morrison knew it all along. She said in a 1993 interview with the *Paris Review*: "I think of beauty as an absolute necessity. I don't think it's a privilege or an indulgence. It's not even a quest. I think it's almost like knowledge, which is to say, it's what we were born for."

• • •

For most of human history, cleaning oneself was more about spirituality and ritual than about any modern notion of health or beauty. In the fifteenth century, the Aztecs cut enormous pools into the sides of mountains for rites of purification. Midwives invoked the water goddess Chalchiuhtlicue as they washed infants, imploring them:

> *Approach thy mother Chalchiuhtlicue. . . . May she receive thee! May she wash thee! May she remove, may she transfer, the filthiness which thou hast taken from thy mother,*

from thy father! May she cleanse thy heart! May she make it fine, good! May she give thee fine, good conduct!

Even slaves whom the Aztecs were preparing for sacrifice were purified with holy water. Ancient Egyptians would dress as gods and ritualistically wash their dead to facilitate the transition to the afterlife.

Hippocrates, the Greek physician in whose name doctors take an oath to this day, advocated baths as something slightly closer to a health-oriented practice. But his interest had nothing to do with removing bacteria (the concept of which would have blown his mind—turned it into fire and smoke). For him, bathing was about a combination of cold and hot immersion that was meant to balance the humors. Warmth was believed to help a variety of ailments, including headaches and inability to urinate. Cold baths were prescribed for joint pain. The processes were fundamentally more about exposure to elements than eradication of any particular source of disease.

These practices famously came together in ancient Roman baths. Citizens of all classes would commune in public facilities designed as much for socializing and leisure as for bathing. Many bathhouses featured open courtyards where visitors could exercise, surrounded by chambers containing a hot pool (*caldarium*), a lukewarm pool (*tepidarium*), as well as a cold pool (*frigidarium*). Some also had entertainment spaces, libraries, vendors selling food and drinks, and prostitutes.

Romans in the baths would sometimes rub themselves with oil and scrape off grime or mud with a device shaped like a sickle. But any hygienic benefit of the baths would have been serendipitous. The water in the pools was far from sterile to start with—some contemporary writings suggest it was sourced from public troughs—and healthy and infirm bathers soaked side by side. The philosopher Celsus prescribed baths for myriad conditions, including inflamed intestines, small pustules, and diarrhea. Without modern chlorination or circulation systems, pools were likely filmed with scum, a layer of dirt and sweat and oil shimmering on the water's surface.

Taken together with the indolence and the nudity, the scene made the baths a flashpoint of the culture wars of the time. The philosopher Seneca saw the decadent facilities springing up in his hometown as evidence of its moral decline. Bathing would also be discouraged by the early Christian church.

Jewish law around the time of Jesus emphasized the importance of the purity of one's body by way of dietary and hygienic ordinance. Ancient Hebrews had laws about washing your hands before and after a meal, and your hands and feet before entering the Temple. A rabbinical saying that translates to "physical cleanliness leads to spiritual purity" has been cited as the origin of the cleanliness-godliness adjacency aphorism.

Early Christians began to move away from the ethos of regimen and restriction, many leaving behind strict Jewish laws about forbidden foods, circumcision, and keeping the Sabbath. Their messiah, Jesus, was a relative minimalist when it came to ritual purification. Artists would later render his skin and hair free of grime or knotting, but, like so many people who develop loyal followings, Jesus was also vocally unconcerned about his personal aesthetic. In the Gospel of Matthew, he scolded those who put religious ceremony before inner purity: "Cleanse first that which is within the cup and platter, that the outside of them may be clean also." Elsewhere in the New Testament, he and his disciples shocked the Pharisees by eating bread without first washing their hands. In the fourth century, Saint Jerome ordained: "He that is once washed in Christ needeth not wash again."

Apart from the symbolic practice of baptism, Christianity was an outlier among major world faiths in terms of bodily cleansing. It stands apart for its lack of bathing or hygiene requirements. Islam, by contrast, prescribes ritual washing before prayer five times a day. The need for water at mosques gave Arabic cities reason to build elaborate water systems that the Europeans lacked. In the 920s a Muslim envoy traveling along the Volga River described the Vikings he saw there as "the filthiest of Allah's creatures" as they "do not wash after shitting or peeing, nor after sexual intercourse, and do not wash after eating. They are like wayward donkeys."

Hinduism, too, involves prescient mandates for hygienic practices. Centuries before Western germ theory, people were to wash their hands after defecation. Only the left hand was to be used in attending to such tasks, and only the right for eating. When the Italian traveler Marco Polo visited India in the thirteenth century, he was taken aback by how fastidiously everyone drank water. They all had individual flasks, he mused, and "no one would drink out of another's flask. Nor do they set the flask to their lips." Even more astonishing to him, the people of India regularly bathed.

Polo had been similarly fascinated in China, where he noted, "There is no person who does not frequent the warm bath at least three times in the week, and during the winter daily, if it is in their power. Every man of rank or wealth has one in his house for his own use." Such was not the case in his home of Venice. When the various groups later known as barbarians overthrew Rome, they destroyed many of the aqueducts and baths. A lack of infrastructure combined with the Christians' skeptical stance on hygiene to render the Middle Ages, as they would come to be called, "a thousand years without a bath."

This came to a head in the mid-fourteenth century, when dark, festering lumps began appearing in the groins, armpits, and necks of Europeans. Giovanni Boccaccio's book *The Decameron* describes them as being as big as eggs or apples. Three days after these growths appeared, the person would die. As this "black death" tore through Boccaccio's home city of Florence, he described mothers abandoning their own children, and no respite anywhere from the smell of corpses. In spite of prayers and processions, the disease spread unchecked. Three years later about a third of Europeans would be dead.

The lumps were swollen lymph nodes overwhelmed with cells of the immune system, sent into emergency overdrive by exposure to the plague bacteria. But this process wouldn't be understood for another five hundred years. So Christians blamed Jews, accusing them of spreading poison around every city. Given the choice between being burned alive and being baptized in the name of Jesus, some Jewish prisoners confessed and were cleansed of their supposed sins. Others were not.

A more erudite theory attributed the problem to planetary alignment. The medical faculty of the University of Paris issued a 1348 report to explain why everyone was dying: they wrote that Saturn and Jupiter had unfortunately aligned with Mars, "a malevolent planet, breeding anger and wars." Because Mars was in retrograde, it "attracted many vapors from the earth and the sea which, when mixed with the air, corrupted its substance."

This idea of vapors causing disease was known as miasma.

Though it sounds not unlike our modern ideas of airborne contagion and pollution, miasma was about spiritual contamination. In Paris the doctors warned that "the bodies most likely to take the stamp of this pestilence are those which are hot and moist," but also those bodies "bunged up with evil humors, because the unconsumed waste matter is not being expelled as it should; those following a bad life style, with too much exercise, sex and bathing." Avoiding these horrible vices did not guarantee safety, but they assured panicked citizenry: "Those with dry bodies, purged of waste matter, who adopt a sensible and suitable regimen, will succumb to the pestilence more slowly."

Fearing hot water did not help the already horrendous hygiene situation. When the land to bury corpses in Avignon ran out, the pope declared the river consecrated space. Families heaved their dead into the Rhône with consciences clean. The same cannot be said of the waterways. Everywhere people carried fleas that carried plague, which recurred somewhere in Europe almost every year until the start of the eighteenth century. Officials shuttered bathhouses over concerns that they spread disease. The journalist Katherine Ashenburg recounts that as a result of the panic and lack of understanding of the bacteria, the sixteenth and seventeenth centuries were "among the dirtiest in the history of Europe."

The rates of death did not make a compelling case for city life. One was safer in the country, and there was more work to be had. This shifted with the Industrial Revolution. Before the nineteenth century, major cities would consist of a few hundred thousand people. There were no high-rises, or factories to create the hallmark urban haze that now hangs almost constantly over cities like Los Angeles, Hong Kong, and Delhi.

By 1801, London's population had surpassed one million people. It reached more than two million by 1850. Paris and New York would soon follow, as people poured into cities. They did so faster than infrastructure could be built. Sudden crowding made environments visibly dirty: the unpaved streets would be dusty in summer and muddy the rest of the year, with horse manure everywhere underfoot and coal fires polluting the air. Alleys became cesspools of human feces, and water supplies became choked with waste. These conditions led to outbreaks of transmitted disease that would change the world, and create the field of public health.

In the 1840s, as typhoid and typhus epidemics ravaged Europe's industrial slums, the German physician Rudolf Virchow made the connection between living conditions and disease. His work, still informed by miasma theory, led the American Medical Association to study conditions in the U.S. In 1847, it called for ventilation of toilet areas to allow disease-causing vapors to disperse.

The theory of bad air was called into question in 1854, when physician John Snow traced a London outbreak of cholera to a well. His process of deduction involved detailed maps and questioning sick people in search of common habits or exposures. The method predated Sherlock Holmes and would prove so important that it gave birth to the modern field of epidemiology. Even still, he didn't understand how water could be causing disease. Nor was he taken seriously.

The idea that the pit of human excrement adjacent to the well had contaminated the water with organisms invisible to the naked eye was not only anathema at the time, it would have had massive political implications. The entire city would have had to be overhauled to separate human waste and drinking water. The London government dismissed Snow's findings as a spurious correlation. He would not be vindicated until two decades after his death when, in 1883, German physician Robert Koch saw the cholera-causing microbes under a microscope. Combined with the epidemiology from the London well and subsequent observations, Koch solidified the case that contaminated water was indeed to blame. And if these "germs" could sneak undetected into our water supply and kill us, it stood to reason they could also be behind most any other illness, malady, or temperament.

This new "germ theory" gradually took hold in the public imagination, at the same time that rapid urbanization and population growth compounded the threats of infectious diseases. Combatting and preventing them would become integral to urban planning around the turn of the century, a period sometimes referred to as the "hygiene revolution." It was a direct result of its better-known industrial predecessor. Public health arose as a newly necessary field to promote basic sanitation and hygiene in Europe and the United States. The past political expedience of denying germ theory gave way to urgent investment in preventive infrastructure. Priorities included pathogen-free drinking water, sewage systems, and getting people to wash their hands after defecating (as other people around the world had been doing for millennia). Until these changes, occasional decimation of entire neighborhoods or cities had been accepted as a fact of life. The understanding that this could be prevented was revolutionary.

Ideas about personal hygiene also rocketed to the center of consciousness. A person's *cleanliness* could be taken as a marker of who was or was not dangerous. To appear ungroomed suggested that you could not afford to wash, and that your toilets were the excrement pits in alleys adjacent to your tenement. You may be one of the disease carriers. On the other hand, appearing groomed—with washed clothing, combed hair, and smudge-free skin—was a signal of safety. Though grooming was not a guarantee that a person washed their hands or didn't have fleas—the actual, disease-causing concerns—appearances and hygiene became conflated.

As these concepts of cleanliness and dirtiness became more concretely tied to health and death, respectively, divisive connotations also spread. To appear clean required resources—money and time. Indicators of hygiene became proxies of status, and more was often seen as better. It was no longer sufficient to simply avoid appearing or smelling physically repulsive; a person was to smell actively *good*. Such demonstrations of cleanliness became ever more of a gatekeeping mechanism for certain professions and social circles. The working classes became known as the Great Unwashed. Upward mobility depended on the pit-adjacent workers being able to dress for the non-pit-adjacent jobs they wanted, not the pit-adjacent jobs they had.

In the early 1900s, people in Manhattan's middle and upper classes began to wash in their bedrooms. Even in impoverished tenements, families might bring out a basin once a week and fill it with water to bathe the children on the kitchen floor. This meant carrying buckets of water up multiple flights of stairs and heating it on a woodstove. These were the lengths to which people would go—and some still do—simply to appear "clean."

The concept of hygiene was also deployed more explicitly as a tool of social engineering. Efforts to contain sexually transmitted infections led to what became the "social hygiene" movement, which undertook public-education campaigns to help control syphilis outbreaks during World War I. The movement would later give rise to in-classroom programming for schoolchildren known as sex education. The scientific justification for addressing such matters in the name of hygiene gave a legitimacy to what would previously have been irreconcilably taboo.

Similar mechanisms would also fuel catastrophe, as long-used language of eradication and cleansing was given a new patina of legitimacy by the emerging science of genetics and infectious disease. In Germany in 1895, physician Alfred Ploetz published a book titled *Rassenhygiene* (*Racial Hygiene*) that would lay the groundwork for the eugenics movement in decades to come, and eventually the Holocaust. Ideas of purity and cleansing became the basis of isolationist arguments, drawing on basic assumptions that homogeneity is good and diversity is unnatural or dangerous.

Fear and contempt of the microbial world would play into forces of division from explicit racism to oppressive standards of sexuality. They would also be used to sell soap. The myriad products that fill pharmacy shelves today began to trickle into everyday use not long after running water and bathtubs became common among the working classes a century ago. The newly ubiquitous practice of bathing would create tremendous markets for soap—and an escalating arms race of other cleaning products. When the poor could no longer be identified as the Great Unwashed, the wealthy would need new ways to set themselves apart as the most clean. Capitalism sells nothing so effectively as status. And if a little bit was good, a lot would be better.

III

LATHER

Dr. Bronner's Magic Soaps began as a church.

The transition from nonprofit religious organization to full-time distributor of peppermint-scented soap was so gradual that it slipped Emanuel "Dr." Bronner's mind to ever give up his namesake organization's tax-exempt status. He spent his final years in bankruptcy, battling the IRS for over $1 million in back taxes. But to the end he answered or returned every call that came in to the company.

Dr. Bronner's flagship product is an amber liquid soap in a transparent plastic bottle that you may have seen anywhere from natural grocery stores to Walmart to celebrity Instagram accounts. The iconic blue label is covered all over with tiny, exclamatory text: "Ready to teach the whole Human race the Moral ABC of All-One-God-Faith! For we're All-One or none! ALL-ONE! ALL-ONE! ALL-ONE!" It goes on like this.

This was Emanuel Bronner's gospel. He fled Germany before the Holocaust, and he traveled across the U.S. in the 1950s to spread a message of peace and unity. He would recite his message to passersby on street corners in Los Angeles from the top of a literal soapbox. He offered soap for sale to help him fund his mission. People didn't especially care what he had to say, but they did seem to like the soap. So Bronner started printing his sermon on the soap labels. Eventually people got word of this strange man's peppermint soap, and demand to buy it rose—even though all he'd intended was to use it as a vehicle for his message of love and unity. So it was important to his grandsons—who revitalized the brand and turned it into the ubiquitous product it is today—to keep the label as close as possible to how he wrote it, despite the inherent marketing challenges.

The brand has swept from niche indie markets to major mainstream distribution in recent years. After half a century of relegation to incense-burning hippie shops, Dr. Bronner's products are now prominently displayed in major retailers, from pharmacies to groceries to boutiques in fashionable hipster enclaves, next to high-end beauty products. Over the past two decades, since David Bronner and his brother, Mike, took over the company, sales have grown more than thirtyfold.

The first thing David Bronner does when I meet him is offer to give me a foam bath in a trailer—but not in a weird way. We're in the parking lot of his company's headquarters in Vista, California, where they relocated a few years earlier after outgrowing their previous space. He and his employees bring the shower trailer to mud runs and events like Burning Man as a communal showering experience. David, now CEO of the company, has been going to the festival since before it was cool, and when he took over the company he thought it would be a good place to involve the brand. Though there's no advertising allowed at Burning Man, and no business can technically sponsor it, Dr. Bronner's does put together elaborate displays and interactive exhibits that convey the ethos of the company. They sponsor a "safe space" for people who are having bad psychedelic trips, and also put on performances by their in-house performance-artist troupe. Whom I am fortunate enough to meet.

"Hey, do you like to dance?" one of the bearded men asks me. (As I recall, all the troupe members had long beards, but it's possible they were just scruffy. This is one of those situations where memory fails, because it was early and I was on almost no sleep and had already had a pint from their kombucha keg.)

"No," I say.

They were clearly expecting me to say yes. Someone turns on a boom box, and they all line up in two rows.

"Okay, right on. Well, we'll dance for you."

Eight grown men do a dance for me in the middle of the warehouse. They watch my face for a reaction, and I genuinely appreciate the effort, though it jangles me to be danced upon like this. They smile the whole time, and high-five one another and me when they finish. Then they stand around in a circle and ask me about the skin microbiome. Some of them tell me they also rarely shower, and as I talk about the idea behind the book they act as if their minds are physically imploding at the conceptual awesomeness. It is like being the first human to give pizza to the Ninja Turtles.

The employment of nonshowering Gen-X artists at a soap company might seem paradoxical. But the company's growth has been a result of developing a strong brand, with which their presence is totally consistent. This brand—based on a general vibe of egalitarian activism—has given Dr. Bronner's an edge in the marketplace, especially among corporate-skeptical millennials. Though the product had a loyal following for decades, it only recently began to turn a significant profit.

I decline the public bath because, while it could be fun, I suppose, as part of a group after a Tough Mudder, this would just be me in a parking lot surrounded by public-relations staff who are quite insistent that I have a good time. So I hop in David Bronner's minivan and he drives me around the campus. He also owns a Mercedes that runs on grease as an alternative fuel source, but he uses the van for work—a larger vehicle being technically preferable if it allows carpooling, he explains.

David Bronner doesn't take a salary greater than five times that of his lowest-paid employee. His hair is long in back and vanishing in front, and he is tall and perpetually leaning slightly backward. His vibe is party but respect the mothership we call Earth. He has been likened to Captain Jack Sparrow, but his drug is not alcohol. It's psychedelics. He is pro-legalization and anti–drug war. This would surprise no one who saw him, unless you knew he was the heir and CEO of one of the fastest-growing soap companies on the planet.

As we approach the main entrance, we see a food truck out front that serves tacos that contain meat. He rolls his eyes. Bronner is beyond proud of the fact that his company offers organic, farm-to-table, local, vegan fare for all employees for lunch. It's prepared by a serious chef who, when I toured the kitchen, fed me a spoonful of his farro-squash salad.

"I understand it's not for everyone," Bronner says, his face unmoving as he stares at the taco truck, as if in a mindful exercise of empathy. He parks the minivan and we go into the dining hall, past an enormous mural of his grandfather Emanuel, the original "Dr." of Dr. Bronner's—though he was not a doctor, nor particularly tethered to scientific reality. David and I draw from the kombucha keg as he tries to explain his soap. He's candid about the fact that he barely showers, and that when he does he washes only his armpits, groin, and feet. For him it was never really about the soap, but about having a vehicle for environmental advocacy.

Emanuel Bronner was also a devout minimalist. He included on the famous label that his soap was "18 in 1"—as in, it could be used by everyone for every personal and household need, from bathing to laundering to house cleaning to brushing teeth. It was the opposite of the rest of the soap industry's push to sell multiple products to the same person. The company has only very recently branched out into selling toothpaste and a few other products, creating mild tension with David's vision for growth of a company that's not selling people anything more than they need. "People wanted toothpaste," he explains. Though I'm far from the only one who attests that a single drop of the peppermint soap works verily well.

The spiritual overtones that lead some people to dismiss Bronner's as new-age nonsense aren't actually new at all. If anything, they're a return to the roots of cleanliness as godliness. For all the eccentricities of the company, the Bronner ethos seems more in keeping with most of the historic ideals of cleanliness than any austere claims about science or health ever could be.

• • •

Humans have used soap throughout recorded history and around the world. But when did the product become something that billions of people use multiple times per day—and not simply because they want to, but because they believe they *need* to?

I go to visit a soap historian, a man described to me as the "Godfather of soap." After several phone calls I was invited to see him and his wife at their home in the suburbs of Chicago. I pull up and ring the doorbell, and the twelve-foot wooden door swings open to reveal a tiny white-haired woman: Fortuna Spitz. She smiles and shouts, "Luis!" and her husband, the Godfather himself, lumbers out of his study and waves me into the living room.

"The two people sitting in front of you have done more for bar soaps than anyone in the world," Luis Spitz says gravely. This isn't the sort of thing I'd expect someone to brag about, and any doubt I have will leave me over the course of the four hours they spend walking me through their vast private museum of soap paraphernalia, teaching me the history of soap.

Luis, who is eighty-three years old when I meet him, was educated as a chemical engineer and entered the soap industry with a job in processing at The Dial Corporation. He has represented the Italian Manufacturers of Soap Processing Plants and Packaging Machinery, and he chaired the first World Conference on Soaps and Detergents in 1977. He has edited and contributed to seven soap-related books published by the soap industry, and is currently an independent consultant to soap production and distribution companies. I don't know exactly how to describe what he does now other than that he knows everything about soap, and he also seems to *make* people in this industry.

"I don't think you expected this much stuff!" he says as I scan the soap-advertising paraphernalia and industry memorabilia that fill every free bit of wall and counter space. The Spitzes actually built their house *around* their soap collection. Fortuna serves me apple pie on a soap-themed placemat.

During the afternoon I spend touring the place, I learn that selling soap is, even more than making it, an art. Soap was in fact first marketed using actual art. At the 1893 World's Fair in Chicago, the Pears soap company's approach to getting its name out there was to print "Pears" in an austere way at the bottom of a painting and display it in an exhibition. The upper floor of the Spitzes' home is a gallery of nineteenth-century color prints (chromolithographs) that you would never guess were ads for soap. The most celebrated and reproduced soap chromolithograph is known as "Bubbles," a painting depicting a curly-headed little boy blowing a bubble.

This high-minded, innocent approach to advertising could not last when the soap boom hit. The crowded marketplace meant more-aggressive approaches to distinguishing a product, including slinging mud at others, creating insecurity in consumers, and making claims far beyond what any soap could ever actually do. This was necessary because, in truth, most soaps are nearly chemically identical. By definition, there isn't too much room to change the product—otherwise it's by definition not soap. The basic process of making soap is on the level of high school chemistry, and has been known for centuries.

Soap is made up of the surfactant molecules, or "surface active agents," that result from combining fat and a water-soluble basic compound, or alkali. Fats—whether derived from animals or from plants, like olive or coconut oil—are made up of triglycerides. As the name suggests, this means three fatty acids and a glycerin molecule. When the triglyceride is combined with an alkali such as potassium hydroxide (also called potash) or sodium hydroxide (also called lye), and heat and pressure are applied, the fatty acids break away from the glycerin molecule. The potassium or sodium then binds to the fatty acids, and this is soap.

A surfactant is a simple molecule. It works because one end binds to water, and the other binds to fat—the oils that stick to our skin and don't wash away with water alone. For example, assume your clothing is soiled with mud. Water alone won't remove it. But if you add a surfactant soap to the mix, the surfactant's oil-loving (lipophilic) end is attracted to the oil in the soil and the water-loving (hydrophilic) end is attracted to the water. These opposing forces loosen the mud and suspend it in the water to be rinsed away.

Though no one knows exactly how or when soap was first discovered, Spitz says, apocryphal tales abound. According to a Roman legend, soap was discovered at a place called "Mount Sapo," where they sacrificed animals to the gods. The ritual left behind both ashes and fat from the animal, and when rain came and mixed them together and swept them down the mountain and into the river, the people washing their togas realized it was going way better than before. "What the hell is this, some kind of cursed water?" they all screamed, and fled. (No, they reverse engineered the process and started making soap.)

Since the chemistry of soap is simple, the process was certainly "discovered" in various places, and approaches varied depending on what materials were available. In places around the Mediterranean, olive oil allowed for a high-quality product that could be used regularly. Marseilles, France, emerged as a bastion of soap artistry. Industries popped up in Savona, Italy, and Castile, Spain, which for centuries would remain destinations where people traveled to obtain the soaps of the masters. Though the process of making soap was straightforward, and ingredients all but identical, there was a clear learning curve and distinction between the homemade product and the professional.

Into the late nineteenth century—and later for much of the world—store-bought soap was a luxury good. My own grandfather grew up in a farm town in Indiana where his parents and neighbors wouldn't dream of buying soap. They would make it themselves after slaughtering a pig. They took the skin and cut it into strips and put it into a big cast-iron rendering kettle, and put the kettle over the fire. My grandfather's job was to keep the fire going. The white fat would melt off the skin, and the pig skin strips would curl up and brown in the boiling lard, turning into a delicacy called cracklings, the eating of which he recalled with visceral nostalgia.

The lard was used around the farm for cooking, seasoning pans, treating wounds, keeping tools from rusting, and lubricating stuff. My grandfather said his mother would collect rainwater and mix wood ash and lard with it to make soap. "I got the impression that, when he was little, he was unaware that you could go to the store and buy soap," my dad recalls. If it was indeed for sale at the general store in his small town, he would've been the first generation for whom that was the case. And having lived through the Great Depression, he was and always remained reluctant to pay for anything that he could make for himself.

We still have the hog hair scrapers, hanging hooks, and rendering kettle at the farm. On the same land where my grandfather would find arrowheads in the dirt, American Indians had thrived not long before. Many tribes were known to practice ritualistic cleansing in "sweat lodges," a ceremonial gathering in a sweltering hovel or tent, where sweating was part of a prayerful process of penance and purification. But this was a spiritual cleansing, where the sweating likely did more to alter one's mental state (through mild and sometimes even fatal dehydration) than to actually cleanse the body. Bathing was done in lakes and rivers. Though there is little record of soapmaking, many indigenous people did have access to saponifying plants—like soaproot and soapberry. Even earlier, Aztecs used two vegetable products, the fruit of the copaxocotl (which the dirty Spanish marauders would call "soap tree," probably just before trying to *kill it*) and the root of a plant that would be classified as *Saponaria americana*, for its soapy properties. These were not named arbitrarily. They produce saponins, which appear to be part of a self-defense mechanism. These are surfactants, just like those that are produced during soapmaking. When these and other plants, like agave or yucca, are peeled, pulverized, and mixed aggressively in water, the process will generate suds.

The mildness of the resulting "soap" would be in demand today. Soapberry suds would be much closer to popular modern cleansers like Cetaphil (marketed for "sensitive skin") than the early soaps were. Putting store-bought soap on one's skin was, for most of the history of the commercial product, not regular practice. That's because making the soap required a base, and the cheapest and most readily available was often lye. The resulting product was highly basic, and could parch or even burn the skin.

Like any tool, this early soap had its place. If you were covered in grime or goo that wouldn't come off with water, using some soap might be necessary. Until the late nineteenth century, though, the primary use of soap was laundry. There were "soap makers" in Jamestown in the seventeenth century, but early colonists more often made the same thing themselves out of surplus animal fat and lye, applying it only in cases of extreme filth. Washing regularly was not simply expensive, but also ate away at clothes and skin alike.

The process slowly improved and made soap more tolerable. As some soapmakers began to use a new base, potash, bathing with soap became more common. A method of processing the ash became the first patent in the United States. The one-paragraph document was approved by Thomas Jefferson and signed by George Washington in 1790, giving rise to a patent process that would shape the future of capitalism.

Intellectual property rights would become central to the growth of the soap industry. In Britain monopolies kept manufactured soap scarce, and a soap tax kept it expensive. When Chancellor William Gladstone finally repealed the tax in 1853, the sudden affordability of soap unleashed an industry that would work tirelessly to overturn the idea that bathing was a vaguely sinful luxury. Quite the opposite: it was a necessary element of basic decency. Through the power of marketing and advertising, the industry would redefine the concepts of health, beauty, and cleanliness. Lingering European taboos around washing regularly would totally reverse themselves. Over the course of just a few decades, it would become taboo *not* to.

. . .

I'm riding on the top of a fire truck around Dr. Bronner's headquarters. As the company has grown, they say they are now required, for liability purposes, to tell me to hang on.

The truck is equipped to shoot foam instead of water. As with the shower trailer, the company brings it to festivals and keeps it around the premises as part of the brand experience. It plays loud music, which feels especially out of place in this suburban office park. The publicists are wearing bright red-and-blue jumpsuits reminiscent of Oompa Loompas. It is a jolt back to reality as we swing around to the loading docks where tanker trucks deliver the oils—much of this cargo shipped all the way from Ghana.

Towering garage doors open onto the production floor, which is a stark contrast to the groovy vibe of the rest of the place: pristine and industrial, with enormous stainless-steel tanks for high-pressure saponification towering overhead. There is a hall of fragrance silos thirty feet tall, in all the colors of their corresponding Bronner's labels. A plastic jug labeled "Citric Acid" (added as a preservative) is taller than I am. The centerpiece, where the saponification happens, is called the reactor, a 1,500-gallon tank with a hatch on top that is shut by turning twelve separate bolts and a maritime steering wheel to lock it into place. The enormous vessel is connected to two equally large ones that contain hot and cold water, and to an emergency pressure-release valve that drains into a monstrous "emergency collection tank." Temperatures reach into the thousands of degrees, and apparently the potential for a massive explosion exists. I climb some scaffolding up to the top of the reactor, where the technician tells me not to fall in and laughs. A horrible death scene flashes before my eyes.

The fundamental principles of soap composition and performance apply to all soaps. Apart from fragrances and colors, the main difference among soaps is what kind of fat is used. That depends on what kinds of plants or animals provide the fat. All fat consists of a chain of carbon molecules. Some are fully saturated with hydrogen (saturated fats) and others have empty sites where hydrogen could bind (unsaturated). Both work well, and most soaps contain mixtures of the two. The conventional wisdom is that soaps from unsaturated fats are more effective as cleansers, but more drying. Soaps made with a greater proportion of saturated fats generate more lather.

Dr. Bronner's distinguishes itself by using only organic plant oils. The label adds that the soap's ingredients are ethically sourced and fair trade and do not involve genetically modified (GMO) crops. Before writing this book I had no idea that these terms could honestly apply to soap. But production of palm oil, one of the most commonly used oils in soap, has been a leading driver of deforestation in many equatorial countries. Environmental advocacy groups like Greenpeace regularly call attention to the environmental impact of palm oil in consumer goods. Soap companies that deal in palm oil have also been implicated in human-rights abuses such as child labor by organizations like Amnesty International. The group has implored Unilever, Colgate-Palmolive, and Procter and Gamble, among others, to adhere to what it considers ethical production of palm oil—and for consumers to insist on products that are fair-trade certified. (Some companies have announced changes, but most mainstream products have not met the advocates' standards.)

The issue is a priority for David Bronner. He insists that the thousands of gallons of palm oil he imports every year come from fair-trade farms. The company is also investing in sustainable agriculture practices, particularly in Ghana. The process is still far from ideal, though. The carbon footprint of flying fair-trade palm oil from Ghana to be refined in Amsterdam and made into soap in California—and then shipped around the world in plastic bottles—is, as the company's chief operating officer, Michael Milam, puts it when I ask, "the elephant in the room."

The manufacturing process is common to most soap factories. Saponification and drying are done in one enormous machine (the reactor), and a computer can control the whole thing. In the Dr. Bronner's factory, an LED display the size of a chalkboard shows a grid plotting out the entire floor, with all the levels in every bin, along with their temperatures and pressures. As I peer into the reactor, cylindrical bottles shoot down a conveyor belt to where machines pour in the golden liquid, apply a cap, and slap on a label. The human work consists of checking for defective bottles and clearing jams.

Bar soaps are a much smaller part of their business. At one side of the factory a hot solid goo is extruded before being chopped into bars and stamped with a logo—a process known as finishing. I grab one hot out of the machine, and it's bendable like rubber. Smaller companies often purchase this kind of "raw" soap wholesale in the form of noodles or pellets, then add fragrance, dye, shaping, and packaging. The profit margins are enormous.

Insistence on using certain oils from certain parts of the world for one's soap is a luxury now made affordable for billions of people. Though few consider the transportation costs or source of ingredients, these have always been the central drivers of cost and availability. As much as any medical or public-health imperative, what fueled the nineteenth-century soap boom was the meatpacking industry. The Spitzes live in Chicago because it has historically been the beating heart of soap selling, what they refer to as "the soap capital of the world." Having grown up in the region, all I knew was that when we passed a rendering plant, it smelled like when a spirit flies up your nose and starts feeding on your soul.

As Chicago's stockyards started overflowing with excess lard that would often be thrown away, young entrepreneurs took notice. Where others saw piles of decaying animal fat, they saw the American dream. They flocked to the city to get into the soap business much the same way that forty-niners had gone to California in search of gold, or that tech entrepreneurs go to Silicon Valley today in search of . . . something.

Among these early "soapers" was William Wrigley Jr., who came to Chicago in 1891 to sell the soap his father made in Philadelphia. To help promote the soap, he gave away premiums like baking powder and chewing gum. The latter caught on better than the soap ever did. By 1895, Wrigley's changed its branding from a girl holding a bar of soap to an illustration of Juicy Fruit that read,

"Manufacturers of Chewing Gum." The iconic Wrigley Building and long-cursed Wrigley Field would not be so named were it not for soap.

A more successful soaper was James Kirk, who built a five-story factory near the mouth of the Chicago River covered in soaring signs advertising his company's four soaps: Jap Rose, White Russian, Juvenile, and American Family. This was an early example of segmenting a product by consumer, Spitz explains—selling not one but four soaps by marketing and packaging them as though they were for very specific people and purposes. The Chicago soaper Nathaniel Kellogg Fairbank (who had bought a lard and oil refinery and started making soap only to avoid wasting excess lard) took this to the next level. In a shotgun approach to branding, he created brands that read like a drug dealer's euphemisms: Copco; Clarette; Chicago Family; Ivorette; Mascot; Santa Claus; Gold Dust; Fairy; and Tom, Dick and Harry.

Differentiating these products all came down to marketing. Fairbank published illustrated booklets called "Fairy Tales" that featured poems and innocent puns, like "People with common sense pay but five common cents for a soap with no common scents—that's Fairy Soap."

Pears also got into publishing, printing, and distributing a magazine known as *Pears' Annual*, which contained actual literary works like Charles Dickens's *A Christmas Carol*. Interspersed were ads for Pears soap—including postcard-size inserts that would fall out upon opening the magazine, an early instance of an infuriating practice that continues today.

Eventually dispatches from the world of soap publishing blurred the line between information and advertising. *How to Bring Up a Baby: A Hand Book for Mothers*, for example, was published by

Procter and Gamble in 1906 and distributed for two decades after. Its pages offered the legitimate wisdom of a nurse, intermingling critical information on how to rear and keep a child alive with tips on how to use Ivory Soap. This early version of today's "sponsored content" would come to be a hallmark of the industry, a precursor to the monetization model on which influencers and some digital media companies now rely.

Of the hordes of entrepreneurs who would spring into existence during the soap boom, desperate to make a name for a new bar of soap on the wings of clever marketing and media strategy, two brothers stood apart from the rest. Their name was Lever, and the company they founded would become the world's largest distributor of soap. They built Lever Brothers—now Unilever—not through innovation in the art of soapmaking, but through the unsubtle art of branding. They sold soap as a health product that would save your life.

James Lever's older brother, William, was born in 1851. William gets all the credit for the business they technically started together. William Lever entered his father's grocery business in Lancashire, England, at sixteen. His job was to cut and wrap soap. At the time, people who wanted to purchase soap would have the shopkeeper hack off a hunk from a huge brown slab and buy it by the kilogram. This soap was somewhere between the caustic homemade lye soaps and the luxury toilet soaps of Castile—something that could be used on skin, at least occasionally, as some people were starting to do.

Lever took over the grocery business, and by age thirty-three was already a wealthy man. Listless and feeling he had fully explored the potential of that industry, he wanted to continue growing in business. The Industrial Revolution was underway, and urban areas were pulsating. Lever recognized that the challenges of city life were also an opportunity to create demand. A new "middle class" was being paid and educated well enough to care about emerging concepts of health and hygiene. As cities accumulated tall buildings that blocked out the sun and factories filled the skies with smog, his mind returned to soap. This was a product that could conceivably be in every household.

In 1884, Lever registered the trademark "Sunlight." In an innovative move, he wrapped each bar of his new product in imitation parchment boldly printed with the Sunlight name. At first, Lever didn't even make the soap. That was done by outsourced manufacturers. His role was in branding and selling it. And this he did with the fire of a thousand suns.

"Lever didn't advertise so much as paint the world with his brand," Spitz recounts. He commissioned famous illustrators to design ads, hung Sunlight plates in railway stations, plastered colorful posters all over the city, launched a newspaper called *Sunlight Almanac*, and distributed puzzles, pamphlets, and a book called *Sunlight Year Book* that contained health advice (hint: use more Sunlight soap).

All of this worked. Demand for Sunlight soon outpaced what Lever could meet by outsourcing production, and he eventually built his own soap factory. But even that process was an opportunity to do something grander. He built houses for workers, and ended up creating a whole town, just across the river from Liverpool. He dubbed it Port Sunlight. It opened in 1889 and quickly became the largest soap manufacturing facility in the world. Lever envisioned it as a sort of utopia, a business model he called "prosperity sharing." By providing affordable housing and a tight-knit community built around soapmaking, he believed he could achieve maximal loyalty and productivity—presaging the omnibus campuses of Google and Facebook, where there are so many amenities that leaving work seems, well, silly.

Mechanization played a major role in making soap widely accessible. The 1904 World's Fair in St. Louis debuted a new soap mill, which Colgate & Company purchased to increase production efficiency. The company differentiated its luxury soap, Cashmere Bouquet, by advertising it as a "milled soap." An ad in the *Ladies' Home Journal* explained that "it is 'hard milled' which means that it is put through special pressing and drying processes that give each cake an almost marble firmness. It is not the least bit squadgy. This special hardness is what makes it safe." And so, "used every day, [it] keeps skin young and lovely."

The idea that soap had previously not been safe was based on nothing. And roll mills are used to refine and homogenize soaps, but they have nothing to do with "special pressing and drying." Spitz explains that there is no "soft" or "hard" milling—this was empty marketing jargon from the very beginning—but even now the appearance of "hard milled," "French milled," or "triple milled" on packaging mostly relies on consumers inferring that if words appear on a package in an exclamatory manner, they must mean something good.

The technologies that truly increased the use of soap, more than any hand-washing campaign ever could, were the automatic soap presses and wrapping machines of the 1910s. Not only could soap bars be much more consistently shaped and uniformly packaged, but it could be done much more cheaply. Unlike today, when "small batch" and "artisanal" products are in demand, the idea of a consistent and predictable product was a selling point at the time.

Mass manufacturing drove the cost of bars down and increased the consumer base. Producing at larger scales also raised the barriers to entry: buying all the equipment and employing a large staff meant not everyone could simply get into the business on a whim. To leverage advantages of scale, companies merged into the multinational behemoths they are today. Lever would become Unilever in 1929, when Lever Brothers merged with the Netherlands-based Margarine Union.

As the soap market flooded, producers needed to distinguish their products still further—from those of competitors, and from their own existing lines year over year. This led to the labeling of soaps for ever-more-specific uses or desired outcomes. The idea that some soaps are *beauty* products and some are *health* products, for example—or that some are for men or women or children or dogs or various types of skin—is much less a product of scientific innovation than marketing genius.

• • •

The most storied day in soap history may never have actually happened. As the tale goes, one morning in 1879, an operator at the soap plant owned by William Procter and James Gamble left a soap mixer running during his lunch hour. This led to an airy mix that was lighter than normal. Seeing no reason to waste a viable product, Procter and Gamble sold it as a novel soap that would float.

This was the story, anyway, until 2004, when a company archivist discovered that Gamble's son had written in a notebook years before the supposed accident, "I made floating soap today. I think we'll make all of our stock that way." In any case, customers took a liking to this new, white soap. It floated in water, which made it easy to keep track of in a washbasin. The "accidental" invention sold so well that Procter and Gamble decided to start making it intentionally.

This may also have been a rare case of a product being created before its branding strategy. As the story goes, William Procter's son Harley was in the throes of searching for a name for the product when he had a revelatory moment during a Bible reading in church. Psalms 45:8 read: "All thy garments smell of myrrh, and aloes, and cassia, out of the ivory palaces whereby they have made thee glad."

The next day he christened the soap: Ivory.

While other soaps alluded to purity and godliness through cleanliness, Ivory drew directly from Scripture. Keeping with the theme, Procter and Gamble decided to advertise it as a "pure" soap. The company went to great lengths to have its purity measured. Five universities and independent labs compared Ivory soap to Castile soaps, which were then considered to be the standard of purity. (Many still consider it to be—Dr. Bronner's signature product is marketed as a "pure-Castile soap.") The results showed that Ivory had just 0.11 percent free alkali, 0.28 percent carbonates, and 0.17 percent minerals. Procter and Gamble subtracted that total from 100 and started advertising the soap as "99 44/100% pure," despite the fact that other soaps were almost certainly comparable, and that additional minerals and carbonates weren't necessarily bad things. Sales boomed, as the religious and idealistic overtones combined with the appeal of a white soap in the post-Reconstruction-era United States.

This message was subtle compared to the marketing of some competitors. The most famous product of Chicago's Fairbank soap company was called Gold Dust Washing Powder, and its ads featured illustrations of the "Gold Dust Twins," Goldie and Dustie, two children with pitch-black skin, precocious musculature, extremely white smiles, and exaggerated lips, often sitting in a washbasin or doing household chores. They became the symbol for the Fairbank Company. Magazine ads urged, "Let the Gold Dust Twins do your work." The homage to slavery is not subtle. The product was so popular that Lever Brothers licensed it for national distribution, and eventually purchased the brand in the 1930s. (It is, for obvious reasons, no longer being manufactured. But as I write this, a metal sign that says "Let the Gold Dust Twins Do Your Work" is for sale on eBay for $3,249.95.)

Other advertisements promised clean hands and racial domination in a single product. An 1899 advertisement for Pears soap showed a naval officer washing his hands in a gleaming bathroom, against a backdrop of colonial imagery. "The first step towards lightening The White Man's Burden is through teaching the virtues of cleanliness," the ad read. "Pears' Soap is a potent factor in brightening the dark corners of the earth as civilization advances."

Racist tropes would become more explicit in later Ivory advertisements, too. A promotion from the 1920s told the tale of white children who came across a "savage village" of thatched huts and dark-skinned natives who "believed in *right* to dirt / And smudge-besmearing sin." The young heroes then scrubbed the natives "until all the village smelled like IVORY and rain."

In a move that holds up better to time, Procter and Gamble chose as the product mascot an infant, who would become known as the Ivory Baby. Slogans centered around the theme, and began to veer toward the medicinal: "Good health and pure soap: the simple formula for beautiful skin"; "If you want a baby-clean, baby-smooth skin, use the baby's beauty treatment—Ivory Soap"; "Keep your

BEAUTY on duty! Give your skin Ivory care, Doctor's advise"; "The beauty treatment of ten million babies."

Even though the claims were awkward and nonsensical together, Harley Procter's ultimate success—and the one for which he is most remembered—was simply to combine his two most popular ad campaigns into what Spitz describes as the top marketing slogan of all time: "Approximately 99 44/100% pure; it floats."

This monstrosity of a slogan has its own trademark. It was elegant compared to the acrostics and cryptic parables of the time. Procter and Gamble transformed from a three-person advertising team into a voracious gobbler of brands. By 1890, the company was bringing in a profit of some $500,000. By 2017, profits would top $15 billion.

As the floating soap trend swept the country, Milwaukee's B. J. Johnson Soap Company sought to break into the novelty market. In possession of palm and olive oils, the company named its soap an oily portmanteau: Palmolive. The product was on the market for a decade before a breakthrough in 1911, when a copywriter in a company meeting said he'd heard these were the preferred oils of the legendary beauty Cleopatra.

If Cleopatra was known for any beauty trend, it was milk baths. According to multiple accounts, she used donkey milk. It was long believed to have special antiaging effects. As the ancient Roman skin care guru Pliny the Elder wrote, "It is generally believed that ass milk effaces wrinkles in the face, renders the skin more delicate, and preserves its whiteness."

Nonetheless, the company decided to use the enduring, regal image of Cleopatra in its advertisements, and the campaign led Palmolive to overtake Ivory as the bestselling soap in the world.

Palmolive was so successful that its Milwaukee makers merged with the larger soap company Colgate in 1928. The new company, Colgate-Palmolive, poured even more money into advertising. Their ads ran in magazines like the *Ladies' Home Journal* and *Woman's Home Companion*, with illustrations painted by famous artists. Cleopatra eventually became a generic beautiful woman: the Palmolive Girl.

Beauty and soap fully merged into one idea with the 1924 birth of the slogan: "Keep That Schoolgirl Complexion." This was a time when almost no women were admitted to institutions of higher education, so schoolgirl didn't mean looking like a caffeine-addled grad student. Palmolive's slogan proffered the cleanliness and purity—and impossible standard—of returning to childhood.

By the 1960s, the messaging became more aggressive and less artful: "New! Continental Palmolive Care can help you be younger looking." And though other soaps made medical and health claims, Palmolive was among the first to invoke doctors. A 1943 ad proclaimed: "Doctors prove two out of three women can get more beautiful skin in 14 days." This would morph into, "YOU can have a lovelier complexion in 14 days with Palmolive Soap, Doctors prove!"

Of course, doctors cannot "prove" something like a "lovely complexion." But factuality is not the point. The specificity of the time frame—and the modesty of the promise that only two out of three women would benefit—offered a sense of plausibility that couldn't have come from saying *everyone benefits instantly*. On the back of the tremendous success of Palmolive, Colgate-Palmolive would go on to become the $15.5 billion company it is today.

Procter and Gamble's simultaneously popular beauty soap

Camay was among the first to invoke not just doctors but dermatologists. A 1928 ad explained: "For the first time in history, the greatest dermatologists in America give a scientific approach to a complexion soap." Below that it elaborated, and not in a sarcastic way: "What is a dermatologist?"

The soap industry pioneered the basic principle of "brand management": that each brand within a company is operated as a separate business, even when the products are very similar. Wanting to compete more aggressively in the beauty space against Lux and Palmolive, Procter and Gamble had introduced Camay in 1923—even though the company already had Ivory, the leading beauty soap. At first, Camay sold poorly. One copywriter proposed that fear of real internal competition was holding Camay back. Procter and Gamble experimented with letting the Camay marketers operate as if Ivory was not their friend. Despite the existence of Ivory, Camay became "The Soap of Beautiful Women."

This practice is still taught in business schools today, and it's why Procter and Gamble has ten indistinguishable brands of laundry detergents alone (Gain, Ace, Era, Downy, Dreft, Cheer, Bounce, Tide, Rindex, and Ariel). Left to their own devices, the soap brands started attacking one another and selling their product as the only safe soap. Marketers of Camay went extreme and essentially introduced the idea of a product cleanse. It implied that all the other brands were toxic, or could not be trusted. A full-page ad featuring a young woman in a bridal gown instructed women to "invite romance with a skin that's lovely! Go on the Camay Mild Soap Diet!" Women were urged to buy three bars ("cakes") of Camay and for thirty days "let no other soap touch your skin."

Though the campaign stopped short of saying "Ivory soap will render you unmarriageable," the message was clear. Even within a company, selling soap was war.

• • •

The village of Port Sunlight is a museum now. For almost a hundred years, the nine hundred houses were filled exclusively with employees of Lever Brothers (and later Unilever). In the 1980s, houses began to be sold to private owners, and though Unilever maintains a research facility there for "personal care products," what was once an empire of soap is now a leading producer of shampoos, deodorants, and products like Axe body spray.

Over the past decade, sales in the bar soap market have been dropping. CNN coverage of the decline cited young people finding bar soap to be "gross." Spitz blames the rise of "shower gels" and liquid soaps, which he describes with audible disgust. Not only are the plastic bottles wasteful (compared to the paper wrapping of many bar soaps), but liquid soaps are also heavy and transporting them is environmentally inefficient. Plus many "liquid soaps" are not true soaps at all, but detergents—a class of synthetic compounds that can mimic the actions of soap, developed by the U.S. Army during the lard shortages of World War II.

To consumers this distinction may be moot, but it is absolutely not to artisanal soapers or to the industry. Detergents were the most consequential technological advancement in the world of cleaning since the soap industry's inception. They are often made from petroleum, which means they can be produced even in places without access to animal fat or fine plant oils. They can be mixed into a wider range of formulas than soap, giving them an edge in laundry and dishwashing. They also appear in the majority of shampoos, body washes, and liquid soaps.

For as much as internal competition built the soap industry and explained its success, the same forces also led it to undermine its core message—that its product was truly necessary. In order to differentiate their products and expand into new product lines year over year, soap companies had to sell the idea that soap was insufficient on its own—or that its effects had to be undone by yet more products. Shampoo alone, for example, would leave your hair dry and brittle. So you also needed conditioner. Soap will leave your skin dry and brittle. So you also need lotion or moisturizer.

This trend reached a critical inflection point in 1957. In an attempt to differentiate itself from its many competitors, Lever Brothers introduced a product called Dove with the slogans: "Looks like a soap, it's used like a soap—but it is not a soap" and: "Dove won't dry your skin like soap."

Dove was indeed not soap—at least not a "pure" soap. It contained (and still does) an emollient cream, or moisturizer. This decreases its cleansing soapiness but also decreases its propensity for drying skin. That is, the product becomes closer to nothing. The addition of an emollient brought the soap's pH down to neutral, so it did not have the same drying effect on the outer acid layers of the skin as a typical soap.

Though it may not have been recognized at the time, this planted a seed in the minds of consumers that soap was not necessarily good or sufficient. There was more out there to be applied to skin in search of this elusive concept of cleanliness than just soap and water. In time this tension, inflicted by the soap sellers on themselves, would birth the modern-day rebellion of indie brands and the vast empire known as skin care.

Still, nothing would challenge the dominance of soap like the changing landscape of media. From the beginning, the key to success in the soap industry was to dominate whatever new media platform presented itself. The first commercial radio broadcast in America was in 1920, covering the election of President Warren Harding. By the following year, there were hundreds of radio stations. Owners realized that sponsored programming was the way to make a business: get these talking boxes into everyone's homes and fill their ears with ads for products.

People turned out to want their ears filled. Families were soon spending evenings gathered around the living-room radio. And when the stations wanted advertisers, they needed to look no further than the booming soap companies, who were eager to further solidify demand and ingrain their product as part of a wholesome, healthy, sophisticated lifestyle. But the soap companies didn't just run ads; they changed the medium itself.

The soap industry used focus groups to determine that its target market (housewives, the primary buyers of household goods) liked to be entertained by radio—not instructed by it. In 1927, Colgate-Palmolive funded *The Palmolive Hour*, a musical-comedy program interspersed with pitches for soap. The program's success led to *Clara, Lu, 'n Em*, a radio show sponsored by Super Suds Fast Dissolving Soap Beads, which promised three "gossiping housewives" chatting about relatable things every night of the week. In addition to delivering proper entertainment to a target demographic, the women also conveniently mentioned Colgate-Palmolive products. The show was so popular it would go on to become the first daytime serial on a radio network.

Not to be outdone, in 1933 Procter and Gamble took to the air to sell Oxydol Granular Laundry Detergent with *Oxydol's Own*

Ma Perkins. The protagonist, Ma Perkins, was a widow in financial straits—the sort of woman who didn't have the time or energy to fuss about laundry. Luckily there was a detergent product that made her life manageable: Oxydol. Though the program may not have been particularly artistically ambitious, informative, dramatic, or funny, it stayed on the air for twenty-seven years. This meant the show met the criterion for good American radio: it sold ads.

Lever Brothers and other soapers created similarly long-running, simple, loyalty-based shows that would eventually come to be known as soap operas. The most enduring, *The Guiding Light*, started in 1937 as a radio show by an ill-named soap company called Duz ("Duz does everything"). It was in the right place at the right time during the rise of moving pictures, and would become the longest-running scripted television show in history.

Before "talkies," people were usually famous because they had done things in the world, like inventing the aircraft or leading a country into or out of war. There were musicians and actors, but their faces were not ubiquitous, and their lives were not tracked in minute detail to the point that they would have the power or credibility to drive many people to buy a certain kind of soap. Now, with their moving faces looming over awed crowds, movie stars would become the original influencers.

Film and television also fueled an obsession with skin more generally. Low-definition cameras paired with makeup and lighting made actors appear preternaturally smooth and infantile, and the tricks behind the practice were not common knowledge. The stars of the screen seemed to truly be either a genetically superior species or in possession of carnal truths about body maintenance that the public could only hope for hints of. When a "testimonial" disclosed some element of a skin regimen that could explain a star's appearance, serious sales ensued. Hordes of stars agreed to say they used Lux Soap, for example, and to have their names and images used in ads that promised "9 out of 10 screen stars use Lux Toilet Soap." Lever never even paid them, and the practice being so new, the stars apparently didn't think to ask.

The meaning of "soap opera" eventually morphed to define a tonality and a familiar set of hyperbolic plot devices, and the term remains in use despite the distancing of soap companies from their creation. When *Guiding Light* was canceled in 2009—an end marked by jokes like Stephen Colbert's hefting an ostensible "DVD boxed set" onto his desk that was about six feet long—*The New York Times* and the BBC heralded it as the end of an era. Other soap operas had also been declining in viewership since the target audience—housewives who could reliably tune in day after day to keep up with the inordinately complex plot lines—was shrinking. The soaps were replaced by game shows and talk shows that can be watched on a periodic basis, ideally optimized to exist as short clips on our phones, to be consumed in occasional moments before another notification pops up and sucks our attention to something new.

Procter and Gamble still owned *Guiding Light* at its cancellation and said it was looking for a new home for the show, but never found one. Not only are people no longer watching soaps, they are cutting TV cables altogether. With Gen X and millennials simultaneously engaging in KonMari- or #vanlife-inspired downsizing, their environmentally conscious minimalism has led to a rejection of many products—and upscaling and hyperawareness of sourcing and quality of others.

Those include the self-care products related to skin. Mass-market bar soap sales are in decline, while "indie" soap brands and skin care companies are infused with venture capital funding and selling as fast as they can fill everyone's feeds on Instagram. The wrenching of the new generation's attention away from TV screens and billboards (just as attention was wrenched from radios a generation before, and trolley-car signs a generation before, and commissioned paintings a generation before) may be the challenge that the soap industry cannot conquer. The monopoly on attention can no longer be so reliably purchased by corporate behemoths. This has opened a new entryway for start-ups, gurus, and influencers to guide consumers to their products.

IV

GLOW

A line of excited young people is winding along a sidewalk just off of Canal Street, outside what I might have assumed to be a night club. Except that it's 6:00 p.m. on a Tuesday, and the crowd is not surly, gelled men but almost entirely women who look to be, on average, around eighteen years old. They seem like the cool ones at their high schools.

The group is waiting to breach the barriers of the new flagship brick-and-mortar store of one of the world's fastest-growing skin care companies, Glossier. The bouncers controlling the line are also young women, uniformed in pink sweatshirts, lifting the velvet rope and marshaling small herds of customers down the hall and into a four-person elevator. Notebook and pen in hand, I have never been more of an interloper.

The customers here possess the sort of skin that advertisements have long taught consumers to aspire to. They have, as it were, "that schoolgirl complexion." They do not appear to use a lot of makeup—that is part of the Glossier ethos, a "natural" look that pushes back against the elaborate cosmetic trends of generations past. Glossier uses the slogan "Skin first. Makeup second." If makeup is about covering the skin, what Glossier is selling is, in theory, the empowerment of showing skin off. The models in the company's ads look like they have just risen from a long and restful sleep and purchased a green smoothie. Their faces, which have a light sheen, bear no evidence of life's hardships. Nor do they seem to have toiled to look so flawless. They simply, to quote Beyoncé, woke up like this.

Exiting the elevator and stepping into the store—which feels like an art installation, with bright white light pouring in from all directions, somehow spare and yet also overwhelming—we are met by pristine rows of white pedestals bearing still more-pristine white tubes and vials of products, from cleansers and serums to lip balms and other skin care "essentials." Ever-present mirrors provide opportunities to compare ourselves to the glowing models in the pictures around us. The product labels are chemistry-conscious: "pH-balanced," "paraben-free," "alpha-hydroxy acid."

This buzzing, heavenly shrine to skin is a spectacular collision of the worlds of beauty and health. The soap industry first invoked dermatology a century ago as a source of legitimacy. The all-encompassing power of skin care now seems poised to subsume dermatology almost entirely.

Glossier sprung from the brain of Emily Weiss, who may need no introduction here. But just in case, she entered the industry as an intern at *Teen Vogue* and launched a blog about skin, beauty, and wellness called *Into the Gloss* in 2010. She built a loyal community by interviewing women about their skin care and makeup routines. The stated goal was to give people a platform to talk about what beauty meant to them, rather than what big corporations made them think they needed. As Weiss has said, she started the site because she "became more and more aware of how flawed the traditional beauty paradigm is. It has historically been an industry based on experts telling you, the customer, what you should or shouldn't be using on your face."

Weiss launched her own line of four products in 2014, at age twenty-nine. Given her prominent blog, the line was well positioned to develop the "cult" following it did—the descriptor, as with any trend on the scale Glossier would achieve, being only somewhat inappropriate. The initial line included a face mist and moisturizer, but it really exploded with a product that has become millennial-handbag canon, an eyebrow wax called Boy Brow. It re-creates the effect of the oils that are secreted by hair follicles when we do not wash them off. Its popularity brought millions of people into the Glossier consumer base.

Weiss has been described as the millennial Estée Lauder, an homage to the entrepreneur who started off mixing homemade, age-defying facial creams and selling them to women as "jars of hope." Lauder expanded her product line in the 1940s and eventually franchised a cosmetics brand that would land her among Time's "twenty most successful businesspeople of the twentieth century" (she was the only woman on the list). Lauder's breakout product, introduced in 1953, was a scented oil called Youth Dew.

Glossier—whose mission statement reads in part, "Glowy, dewy skin is our thing"—is now valued at more than a billion dollars. What started as a blog taking on the traditional beauty paradigm is now a company with forty different products, including fragrance and body lotion. In 2017, New York's governor Andrew Cuomo proudly announced that Glossier would move into a 26,000-square-foot office in SoHo, adding 282 new jobs and receiving $3 million in tax credits. While most of the company's products are sold online, Glossier has now opened two flagship storefronts—one in Los Angeles, and the one in New York where I find myself.

I'm accompanied at Glossier by my friend Leah Finnegan, who writes about consumerism, online culture, and feminism, among other things. She explains that Weiss's story of entrepreneurship in an industry long dominated by male CEOs is part of the company's appeal. At an event hosted by *The Atlantic* in 2018, Weiss spoke about being a female entrepreneur in a traditionally male space. As she explained how she has grown the company to meet demand—moving beyond the "natural" look to cater as well to consumers who want to look "like a Kardashian"—she couched the expansion as allowing women to "make their own choices."

Even as she is now among the leading entrepreneurs and tastemakers in skin care, Weiss continues to highlight her outsider status: "For so long [the beauty industry] has been . . . held tightly by a handful of conglomerates who are multi-hundred-billion-dollar companies," she said at the *Atlantic* event. "Luckily we're in an era of social media and personal expression where everyone can be her own expert."

Though, of course, when everyone is an expert, no one is.

Leah sees this empowerment pitch as an illusion. "Of course I'm all for female CEOs, but do we really need to be told to do more skin care? Is that the best way to use your power and influence?" She counters Weiss's claims of being a champion of women by pointing out that she is also selling them extreme, unattainable standards of beauty. The fact that these products and standards are coming from a woman doesn't make them good. "The standards themselves are the issue. It's authoritarianism."

She's sort of joking and sort of not. The city truly is choked with stores that sell products for skin, from bodegas to pharmacies to department stores, even if few have lines outside them. There are billboards and physically impossible mannequins and glossy magazine covers creating and perpetuating ideas of good and bad, right and wrong. Corporations that carefully craft these messages also celebrate themselves whenever they deviate even slightly and include someone who isn't extremely skinny, or who is over forty, or whose skin is less than wrinkle-free and perfectly toned.

Wandering the New York store, trying not to look too out of place, I'm drawn to a product called "Invisible Shield." It turns out to be a sunscreen of SPF 35 that costs $25 for one ounce. It does not promise to be more than a sunscreen. Yet seeing it there, I want it. I feel that I would be somehow better—that I might somehow belong in this place with this crowd—if I were to open it right there and slather it on my face. Even if I simply had it in my pocket.

The products in the Glossier store are beautifully packaged but surprisingly standard in content. The popular acne treatment, a "zit stick," contains the topical antibiotic benzoyl peroxide. This is the most common ingredient in over-the-counter acne treatments. It is sold in some form by most every skin care, cosmetics, and drugstore brand. Glossier's product is $14 for just over one tenth of one ounce. I pull out my phone and see that Walmart sells a stick containing 1.5 ounces (nearly fifteen times as much) for $5.

Like the soap empires and beauty brands that came before it, Glossier is a story of winning trust by winning the newest media. It is also, Leah explains, about people "wanting to become Emily Weiss by buying her products." Her face is synonymous with the brand, which is suffused with jet-setting urban sophistication and carving one's own route to financial independence. Weiss, whose company has more than 2.5 million followers on Instagram, has harnessed the potential of communication platforms that would've made William Lever foam at the mouth. One fashion magazine called her "a pioneer in translating content into commerce."

Unlike the celebrity endorsements of the past century, Weiss also built an expansive network of "reps" in the wellness and skin influencer space who don't necessarily have large audiences. But they do have loyal ones, mostly on Instagram. These reps get commissions and store credit for helping sell Glossier products. As the journalist Cheryl Wischhover recounts, Weiss capitalized on "that age-old adage that the most trusted recommendation comes from a close friend."

But the first mistake in the wellness industry is thinking a professional influencer is your friend. An influencer is a person who explicitly wants your attention because they want to monetize it. Nonetheless, they are popular with the kids. At a recent wedding I talked with a thirteen-year-old whose iPhone case was plastered with a big Glossier sticker. I asked if she was an influencer, and she seemed embarrassed to have to answer no. Trying to turn things around, I said that everyone influences people in their own way, and she half smiled and went back to scrolling on her phone.

Venturing into the heart of the Glossier experience, I feel the acute anxiousness of not belonging. I ask Leah where she gets skin products, and she says, without hesitation, "CVS!" So I suggest we go, also newly curious about what exactly makes the myriad skin products at the myriad drugstores on almost every other street corner so different from those that warrant a line of cool teens outside of Glossier. We head for the line to get back on the elevator.

As we walk through SoHo we pass Credo Beauty, an all-glass storefront whose panes advertise "The Largest, Safest, Real-est Clean Beauty," which sells "hair, body, skin care, makeup." The line is among the many that invoke the word "clean" in the now-common, meta way: not to describe what the products do but to describe what they *are*.

"Clean beauty" is a movement that sometimes refers to minimal environmental impact but more often refers to an undefined idea of purity—just as the word always has. The label has also come to replace "natural," another term with more of a vibe than any standardized meaning. Critics have pointed out the flaws in using "natural" as a synonym for "good": Rattlesnake venom is natural, as are hurricanes. Toilets are not.

Among the critics of the term "natural" has been one of its leading purveyors. In a 2016 announcement on the company's website, Gwyneth Paltrow's wellness-business empire Goop lamented the personal care product industry as "essentially unregulated" and riddled with products containing toxic chemicals. "Because it's a free-for-all, companies can use whatever adjectives they'd like when it comes to marketing and 'greenwashing' their products—*natural*, *green*, [and] *eco* literally have no enforceable definition. In other words, what is touted on the front in no way needs to match what lives on the ingredient label on the back. At Goop, we are creating a new standard of beauty, one that we simply call 'clean.'"

Today Goop continues to distribute all manner of products labeled "natural"—a search of the company website finds 762 posts and products for sale, from Premium Natural Toothpaste to Natural Pilates and an All Natural Scented Eye Pad. But Paltrow has also been a pioneer in selling the same vaguely puritanical ethos under the term "clean." Since 2016 the company has launched an entire line of Goop-branded "clean" skin care products, as well as a "clean" cookbook and even products that promise "clean sleeping."

The marketing approach behind products like these represents a new, transcendent level of purity seeking: not only must one clean oneself but it must be done by way of products and practices that are themselves clean.

The concept is creeping even into the products at workaday drugstore chains. These stores have long played a part in conflating beauty, health, and well-being by maintaining entire aisles of soaps, shampoos, body washes, lotions, and other skin products. At CVS, for example, Leah and I find too many versions of benzoyl peroxide to count. Alongside cheap generic versions are luxury products, too. One line is called La Roche-Posay Laboratoire Dermatologique, whose "body care" products are "recommended by dermatologists worldwide" and "clinically shown to reduce dry, rough skin." There are soaps and alcohols for removing oils, and moisturizers for adding them. There are dozens of sunscreens.

It may be because of the surplus of choices, rather than despite it, that brands like Glossier succeed. The marketplace is so full of products that choice becomes exhausting. Glossier offers curation. If a product is at this amazing store, and Emily Weiss uses it, then it must be good. Or at least safe.

Dispirited and sweaty, Leah and I ride the F train back to Brooklyn. It's one of those summer commutes when people are literally shoulder to shoulder in a steel box that feels about 90 degrees and one is inevitably reminded that people are sacks of metabolically active organic matter. I'm truly grateful to be sandwiched by people with effective hygiene regimens. The cars are plastered with ads for a line of hair, skin, and nail vitamins marketed by the supermodel Heidi Klum, who smiles down at us with blinding white teeth. Her skin is dewy, her hair is inexplicably blowing while her dress is not, and the brand she is selling is named Perfectil.

• • •

The slender hipster cowboy behind the bar is wearing a several-gallon hat and a bolo tie. I sit down and he pours me an unsolicited shot from a brown bottle that looks to contain whiskey. But the liquid comes out thick and gooey, like syrup. If this were the old West, I'd have probably shot the man on the spot. But this is the Indie Beauty Expo, the world's largest annual gathering of independent beauty brands, so I hold my tongue and let him explain himself.

What he poured from the bottle is, he says with a grin, soap. For men.

The proprietor seems desperately happy to have someone to talk to. His products appear to be of little interest to the crowd of mostly women streaming past us to other booths.

His brand is called 18.21 Man Made. The numbers are an homage to the Eighteenth and Twenty-First Amendments, which respectively prohibited and unprohibited alcohol. I gather the name was not meant to make analytical sense but to appeal to the basic instincts of free association: things men are assumed to like and feel comfortable consuming. The word "man" is right there in the brand name. It is also in the tagline: "Premium grooming provisions that the society of Men will take pride in owning." The whiskey soap bottles look fully drinkable.

The whiskey theme is common among skin care products marketed to men. Whole Foods sells a line of soaps called Dear Clark in brown bottles with a red melted wax logo that looks eerily like Maker's Mark. One store I visited in Minneapolis sold a brand of black-bottled soaps and moisturizers called Every Man Jack, which appeared in an entire section of personal care products called "All

Things Manly." Almost everything in the men's skin care section at the pharmacy is black, brown, or gray. Instead of lavender and hibiscus, the scents are things like "mountain spice" and "fastballs and fisticuffs."

Differences between these products and their feminine counterparts—which reliably cost more—tend to be fragrance, color, and packaging. These things seem to matter now as much as ever. Skin care for men is a growing market—up 7 percent from 2018 to 2019, and expected to reach $166 billion by 2022—but it remains a novelty in the overall industry. The 18–22 demographic has also shown unprecedented interest in gender-neutral products, according to a 2019 market-research report. But short of offering any truly new innovation, sellers often try to break in by defining and articulating who their products are *for*. Specificity is key. If you are selling a product that is for everyone, you are selling a product for no one.

Another male proprietor I see at the Indie Beauty Expo is a Paleo enthusiast standing on an animal pelt, who justifies using body cream because he needs it for his "torn-up CrossFit hands." It turns out he's a former executive at the Food Network. The pitch for his company, Primal Derma, whose logo is a cave painting–style cow, is that their skin products are "Paleo" because they are made with beef fat.

The expo I'm attending is being held in Lower Manhattan, in a convention center nestled amid public housing high-rises. Every year, up-and-coming skin care entrepreneurs gather at the expo to pitch distributors, to network, to find suppliers, and to determine new ways to sell people more skin products. This place is the bleeding edge of what will be seen in stores in years to come. The sellers' eyes are glinting with a bloodlust to overthrow Emily Weiss.

I am accompanied through the 70,000-square-foot expo space by Autumn Henry, the lead esthetician at Exhale, one of New York's highest-end spas. Autumn is an invaluable source of industry knowledge, trends, sales tactics, and the real value of the art, and agreed to be my guide through the expo's vast universe of products.

The sellers at the expo seem to sense that Autumn knows what she is doing, even before she opens her mouth, and that I don't. It's true.

I ask her what makes a brand *indie*.

"Oh, it's just a feeling," Autumn says. Technically the lines between establishment and antiestablishment are blurry. But all the brands present are not widely known or distributed, and most are unfamiliar even to her. Many booths are staffed by the founders of the companies, lots of whom are doing this in their spare time or as a second career, hoping for a big break—getting bought out by a corporate behemoth like the soapers of old, or getting a distribution deal with a national retailer.

"They need a breakthrough product first," Autumn explains. "So a lot of them are putting in all kinds of new ingredients, either chasing whatever the current trend is, or trying to make a new one happen."

As in the soap industry, tremendous pressure to stand out in a crowded field requires companies to hone the art of selling products that none of us would have imagined that we wanted, or needed. This is often done by highlighting or creating some concern over an ingredient or a symptom or a practice that didn't exist the season before.

Here these approaches are laid bare and executed with intensity. The indie brands are necessarily playing things faster and looser than the mainstream brands, drawing out the industry's id.

Once a company breaks out, it's also more likely to draw scrutiny from regulators about claims it makes about its products. But until then, this is a space to take chances.

Many sellers use the strategy of heightening an already existing approach. If people want artisanal, then here they can find products that are made in even *smaller* batches, or that use even *fewer* or *purer* ingredients than the ones you see in *any* store. We pass booth after booth in the expo, beneath airy, glittery decorations hanging from the ceiling. The words "clean," "pure," and "cruelty-free" are ubiquitous. As is "charcoal"—one of the products least intuitively associated with looking clean. Especially if you ask a coal miner. The proprietor of a brand called Sumbody offers me a "fountain of youth stem cell moisturizer." (The "stem cells" are from pumpkins.) Max & Me offers a "Sweet Serenity Mask & Wash" product that "benefits hard to treat skin issues." (The issues are not specified on the sample I receive, but the company's website lists them as acne, rosacea, and couperose. The site also promises that the product, which is mostly clay and honey, "suffuses you with beautiful vibrations.")

Below a neon sign that says "Own Your Beauty," a proprietor is wearing medical scrubs and distributing lotion. The medicinal vibes that pervade the conference hall seem not to be accidental. In the imagery, the pitches, and the products themselves, the light-hearted artistry is also subtly imbued with a sense of life and death.

The relationship of the skin care industry to science is complex, but here I glean some rules of engagement. It is okay to say that products and ingredients are "scientifically proven," and that studies have shown your product is good. But it is not okay to ask where the study was published, or how many people were in it. Unlike "mainstream" science (which many people here and elsewhere distrust or believe has failed them), indie science is less concerned with methodology or statistics. This science is about lived experience and personal expertise. It's the kind of science where a "study" might turn out to mean that everyone at the company tried the product and absolutely *loved it*.

Autumn laughs and rolls her eyes at many of the products, but she also spends her life caring for people's skin because she truly believes in the importance of skin care. These products purport to physically change the functioning of our largest organ. "People experiment with this stuff as if it can't hurt, but they also think it might help," she says. "If it's truly doing anything—if something has the potential to help, it also has the potential to make things worse."

Take the popular product Lotion P50. Made by the French company Biologique Recherche, it is, as beauty blog *Into the Gloss* explains, not a lotion at all but "a French water weight exfoliating toner." Toner is a buzzword that has no agreed-upon meaning; nor does "water weight." The product is an exfoliant—and a chemical one, so it burns off the dead skin cells instead of scraping them off with physical force. I'm told that my descriptions of these processes are less enticing than those in the marketing copy, but this is just the literal truth of what happens. Almost every product is either exfoliating (removing dead skin cells) or "cleansing" (removing oils) or moisturizing (adding oils). Exfoliation is the opposite of what some buyers might expect from a *lotion*, but this requirement of insider knowledge is part of the appeal.

Lotion P50 smells terrible. One person described it to me as a tire fire, another as body odor. It also does not feel good, by most accounts, including my own. *Into the Gloss* cautions readers: "Stinging and redness is par for the course," but ultimately this is worthwhile because it has "a hearty blend of AHAs and BHAs to give you that glow. But what makes P50 particularly special is the mix of sorrell *[sic]*, myrrh extract, myrtle, and onion. (That's where the smell comes from.)"

The original formulation of the product, marketed as P50 1970, is banned in Europe because it includes phenol, also known as carbolic acid, originally used in antimicrobial soaps beginning in the late nineteenth century. The compound causes a burning and then numbing sensation, which may be familiar to some in the UK who remember having their mouth washed out with carbolic soap as punishment. Buyers are advised to consult their doctors if they wish to use P50 1970 while pregnant or breast-feeding. The website for one of the product's authorized retailers also warns that "DUE TO THE NATURE OF P50 IT IS PRONE TO ARRIVE LEAKING FROM SHIPMENT." (It's unclear why.)

It costs $101 for an 8.5-ounce bottle. It is beloved.

Part of the appeal of skin care is that it is explicitly not about rational argument but about artistry, autonomy, enjoyment, and personal expression.

But it can also be stressful. The sheer number of options can keep you from ever feeling truly confident in what you're doing. Autumn hears from clients that they ping-pong between feeling up to date and on trend, and concerned that they've been doing it all wrong or missing something important. Instead of offering calming, grounding benefits, skin care can become a source of constant doubt and uncertainty. Albert Einstein, who lived through the heart of the soap boom, insisted on continuing to use soap for shaving, instead of adding one of the fancy new "shaving creams" to his regimen. He is reported to have said, "Two soaps? That is too complicated!"

Of course, he was not the typical man; he rejected material possessions of all sorts and was so averse to frivolity that he devoted his life to discovering one universal theory of everything. But if the skin care scene of the 1930s was overwhelming for Einstein, he would not fare well today. Navigating claims and knowing how to allocate one's faith and time and money gets more difficult in an increasingly crowded and loosely regulated marketplace.

• • •

As the skin care industry merges into the domain of medicine, some doctors are dismissive. Even open-minded ones tell me they can't keep up with all the products that patients are asking them about, and all the new ingredients. The profession teaches us to be cautious about new treatments until they are proven safe and effective. But many patients no longer find it sufficient for a doctor to answer, "I don't think that's been studied, so I would hold off."

Leslie Baumann tries to be both open-minded and evidence-driven. She founded the University of Miami's Cosmetic Dermatology Research Institute, the first such academic center in the U.S., representing an institutional embrace of the skin care industry. Dr. Baumann is the author of the heavy academic textbook *Cosmeceuticals and Cosmetic Ingredients*, which seeks to guide other dermatologists as they take on roles ever less dictatorial and more responsive—like guides through the product menagerie. She regards it with cautious optimism.

One of the major points of confusion she hears about is retinol. Retinoids (chemicals derived from or related to retinoic acid, also known as vitamin A) have been approved as drugs by the FDA. They are found in some prescription medications. But they are also sold over the counter. They are important signaling molecules that regulate cell growth and replication in the skin and elsewhere. There is some evidence that they actually can "turn on" genes that make the skin produce collagen, and "turn off" genes for the enzyme that breaks down collagen. So if skin "aging" is largely about collagen depletion—collagen being the structural matrix that keeps it looking "firm" and "taut" and not "squadgy"—then this would be the basis of a semiplausible "antiaging" claim.

Topical collagen, by contrast, is useless. Your skin is designed to keep large molecules out, Baumann explains, so it doesn't penetrate. Drinking it also doesn't do anything for your skin. Collagen is broken down by enzymes in the digestive tract just like any protein, not transported whole from your gut to your skin. Even if it were absorbed into your bloodstream, it would first have to be able to wedge its way down into your dermis. It's like if you needed new tires and you put rubber into your gas tank. Yet collagen is everywhere at the Indie Beauty Expo, where I am told that it will firm and plump up and smooth out and "really just bring to life" my skin. Though this is a dagger of a thing to say to a person, since these are not disease-curing claims, they are perfectly legal.

Making new collagen does require vitamin C (ascorbic acid). A person who is deprived of vitamin C for a few months will start to bleed from the eyes and gums as the connective tissue in their blood vessels breaks down, a condition known as scurvy. Baumann tells patients that just eating one of those chewable vitamin C pills is "much, much, much more effective than those expensive collagen drinks."

The definite way to get vitamin C into your body's cells is the less trendy, time-tested option of eating fresh fruits and vegetables. These also contain other elements like fiber that benefit the microbiome. The stomach contains strong acids your skin lacks that are made to absorb nutrients like vitamin C.

There's some evidence that topical vitamin C can change the skin, too. In one study, researchers gave people topical vitamin C and then measured the mRNA of their collagen genes, and found that the genes were turned on, suggesting the person could be producing slightly more collagen than previously. But this approach has not proven any more effective than simply eating vitamin C, and the exact same compound, when mixed into skin products, can become exorbitantly expensive. The extremely popular product C E Ferulic costs $166 for a single ounce. It's made by a company called SkinCeuticals and promises to protect against UV radiation and pollution. The three ingredients are simply stated right on the front of the bottle: vitamins C and E and ferulic acid.

Individually purchased on Amazon, these ingredients would total less than a dollar. Individually buying ingredients has the additional benefit that pure nutritional supplements can be vetted by third parties like the U.S. Pharmacopeia, which certifies that the vitamin in the container is indeed the vitamin on the label. Once something has been mixed into a skin product, there is no such testing process. All the same, users of C E Ferulic pushed back against my suggestion that it might be fun to make at home.

Though many products contain vitamin C, the acid in C E Ferulic is the key to delivering it to your body through the skin. Unless a product has a low enough pH to make it through the skin's acid mantle, your skin will basically carry the product on the outside. While you might like the oily (dewy) look, any antioxidant effect of the nutrients is lost. There's no way to know this from reading a label.

Baumann explains that many ingredients and brands are proxies for the actual product, which is status. Expensive products tend to sell well, she says, not despite their price but because of it. "It's really sad," she said. "I'll have a lady come in with Crème de La Mer and these $600 creams, and she thinks she's doing everything right for her skin, but she's not on a sunscreen and she's not on a retinoid and she's not on vitamin C." The next patient will be someone who comes in and feels guilty that she's not taking better care of her skin because she's busy taking care of her kids. She's only using sunscreen and a little vitamin A. "I laugh because the second lady's doing better for her skin than the first."

I ask Baumann what she has the hardest time dispelling misinformation about, and she says without hesitation, "Peptides." They are a vague class of compounds that are extremely expensive and for which all sorts of claims are made about revitalizing, rejuvenating, and antiaging. Technically they're just fragments of proteins. A protein is a long chain of amino acids, while a peptide is a short chain of amino acids. When you eat proteins, the long chains are digested into shorter chains, called peptides (from *peptos*, Greek for "digested"). There are almost infinitely many different potential lengths and combinations of amino acids, so it's impossible to say that no peptide ever has any effect. But the term is also so broad as to be meaningless. While they represent an enormously profitable line of products, the claims are extremely difficult to prove, and they can interact with other ingredients when mixed into a facial cream or serum. "They also don't penetrate well," says Baumann. "It's really a hoax."

"There is also hype around growth factors," she adds with consternation. These are a broad class of small molecules that cells use to communicate with one another. Their biological functions are critical and complex, and now some growth factors are purposely added to facial creams and serums. This is marketed as though it is a good thing, as if more growth factors simply means more beauty. While this class of molecules serves vital functions in human tissue, each type of molecule works through elaborate signaling cascades and feedback loops, in concert with hundreds or thousands of other signaling molecules. Baumann likens the idea of putting an isolated growth hormone onto your skin to disassembling a football team and expecting the quarterback to be able to play alone.

"And, oh," she goes on, even though I had only asked her to name one thing, "I *hate* those stem cell creams." All of our skin contains stem cells, which keep multiplying to create new skin cells. This is how it's possible to be constantly shedding dead skin and never run out and be sans skin. The idea of putting stem cells on your skin is, ostensibly, that having more stem cells will make your skin even more lively, or something of that nature. Stem cells are often associated with fetuses, and babies have been central to skin product messaging since the early days of soap. But having more stem cells does not make skin better.

Nor is it possible to apply someone else's stem cells to your skin and expect them to work their way down into the basal layer of cells and become part of your own line of stem cells.

Nor is it ethical to actually sell human stem cells. And even if it were, they wouldn't stay alive in some cream on a shelf for months.

The overwhelmingness of these and other ingredient claims may be intentional. They often sound familiar and basically reasonable, but also cryptic enough to make you feel like you're simply ill-equipped to fully understand them. In truth, anyone would be. Consumers are urged to be their own expert, but information asymmetry and minimal lack of regulation of marketing claims make that impossible. It's as if, by design, the consumer is supposed to simply throw up their arms and try the product. No scientific data or explanation is as powerful as one's lived skin care experience. When buyers do have questions about physiology or pharmacology, or which product would be best for certain circumstances, there are few independent sources equipped to give an objective answer. There are, however, swarms of entrepreneurs ready to dispense answers to questions we didn't even know to ask.

• • •

In January 2018, while I was working on this book, *The Outline* published a divisive story called "The Skincare Con." The writer, Krithika Varagur, observed, "It's normal today for people in certain circles to brag about spending most of their paycheck on serums. The latest skin care trends have a reassuring scientific cast: peptides, acids, solutions, and other things with clinical suffixes that are typically sold in small quantities for large amounts of money.

"But all of this is a scam," she continued, arguing that there was very little evidence and a lot of manipulative marketing being used to sell people on the idea that they need to be improved, when in fact beauty standards are culturally determined. Varagur notes that Glossier, with its emphasis on dewiness, now sells a powder to make skin less shiny. "Such is the circle of life in a capitalist society."

Her conclusion was not that skin care was bad or to be avoided, but: "Before you start a militant skin care regimen, it's instructive to think about why you want one and why it seems like an intrinsic good."

The story was promptly and roundly condemned by the internet at large. Here's just a small cross section of the Twitter discourse: "lol just give this piece to someone who has battled acne for the better part of a decade and tell them skin care isn't important" (1.4K likes); "YOU CAN PRY MY SERUMS FROM MY INCREDIBLY SUPPLE HANDS" (1.1K likes); "I'll never forget that one time I spent money on my skin and improved it's texture and appearance and felt more confident and happy. What a waste, I was duped. A terrible decision. I should not be trusted to recognize and address my own needs" (2.9K likes).

Leah was the editor of the story, and she was completely surprised that people were so protective of what she sees as a "clearly predatory industry." Why is it not cool to call it out? "The real issues are the marketing to impose a sense of necessity to consume a lot and look a certain way, the fetishizing of youth, and the targeting of women versus men," she says. It would be less of an issue if men and women were held to the same standards, but ideally neither would be.

Curious to see how the story was received offline, without any alleged effects of social contagion on Twitter that can amplify negative responses, I assigned it in the class I teach on public-health media. The reaction was the same: as one student put it, the story seemed to be telling readers that their own experiences with their own products were misperceptions—essentially making you feel gaslit, like you're being told you can't trust your senses. The class nodded in unison.

The basic issue with any sweeping critique of people's spending on skin care is that the many good and well-liked products are lumped in with predatory scams. To suggest that consumers are simply gullible—rendered unthinking consumerists by their own vanity—is to blame the individuals for systemic failures. If skin care devotees were forgoing excellent, safe, easily accessible skin health maintenance regimens and medical treatments that had been thoroughly tested and scientifically proven in favor of some mysterious serum they heard about in an Instagram ad, that might be cause for concern among doctors. But many people come to the skin care space because the medical establishment has failed to address their concerns, and the regulatory apparatus has failed to oversee marketing and advertising such that consumers have little way of knowing what to believe. The skin care industry promises hope and a sense of control. The entirety of human history has shown our inclination to abandon skepticism on the off chance that a product or creed or practice delivers on such a promise.

That was the case with Maya Dusenbery, a journalist who spent much of her life with serious acne. Her dermatologists had her try all the various prescriptions, including astringents to dry the skin, oral and topical antibiotics, and eventually heavier-hitting medications like spironolactone and Accutane. The latter is sometimes known as "the drug of last resort," because it has been linked to serious side effects like suicidality and inflammatory bowel disease. She went on birth control to try to modulate whatever hormonal imbalances might be fueling the acne. "If traditional medicine would prescribe it, I'd try it," she told me.

Nothing worked. Then, at twenty-six, just two weeks after starting another round of oral antibiotics, she developed pain in her joints that made it almost impossible to move. She was diagnosed with rheumatoid arthritis, an autoimmune condition often described as the body attacking its own joints.

Still trusting that doctors would figure out the best treatment, Dusenbery started seeing rheumatologists who put her on medications to suppress her immune system—including methotrexate, a drug used in chemotherapy. It helped, but she started losing her hair. She had to get tested every month to make sure the drug wasn't destroying her liver.

Acne is one of the most common reasons that antibiotics are prescribed. Some patients take them for months or even years, despite a very low likelihood that they would be of any benefit and clear evidence that overuse is dangerous. That sort of chronic, unnecessary use of antibiotics renders them ineffective in cases when they're actually necessary. It also disrupts our gut and skin microbes. Such changes are clearly linked to changes in the functioning of the immune system. They seem to play a part in causing and exacerbating autoimmune disease. Maya started to suspect that the oral antibiotics she'd been taking had something to do with her developing rheumatoid arthritis.

So she started looking for other remedies for her acne. She scoured the internet. She tried skin brushing (a popular trend where the skin is brushed like the hair, ostensibly, among other purported benefits, to stimulate the immune system). She spent $90 on "little botanical things"; she tried oil cleansing, oral and topical vitamins, and myriad other trends.

"It's not like people are rejecting proven treatments. These are desperate patients. I know that if you're sick, you'll try anything to get better," Dusenbery told me. "I was not inclined to alternative treatments at all, but as soon as I was sick I was like, yeah, of course I'll try all of this crazy shit. It shifts your calculus."

Occasionally these things made small differences, and her skin vacillated between mildly angry and ambivalent. But ultimately she found that the most intriguing alternative approach was to do less—much less.

In her quest to self-empowerment, she came across one of the skin care internet's favorite terms: something called the "acid mantle." Though the term is controversial—some people think it's a bigger deal than others do—it has its basis in fact. The compounds we carry on our skin are oily, and so, acidic. On the pH scale (where 7 is neutral), skin tends to be around 5. Referring to this as an "acid mantle" dates to a century-old German paper by the dermatologist Alfred Marchionini and colleagues. In "The Acid Mantle of the Skin and Defense Against Bacteria," he conceptualized a film on the surface of the skin that helps to protect the skin from invasive microbes.

If the acidity does this, it does so by maintaining a rich diversity of nondangerous microbes. Acidity is the normal state of the ecosystem of the skin, harboring the microbes that help us live. When the pH of an environment shifts, so do microbial populations. It is these imbalances—less often than any particular *invasion*—that are associated with disease.

This pH-driven model of skin health does not bode well for soaps. A soap, by definition, has a highly basic pH of 10.3. This is by design. The less basic a soap is, the less well it binds to the oils we seek to remove. Dove has a pH of 7, because of the addition of the emollient. This makes it less drying. In other words, it is less able to bind and remove oils. In other words, it is less good at its job. It is the nonalcoholic beer of soaps.

Learning all this, Dusenbery came to conceptualize the problem as one of stripping her skin with all the products meant to make it less oily. And as she cleaned more aggressively it only seemed to get oilier faster. She had been dutifully battling her acne, but she was becoming convinced that the battle was the problem. Eventually she threw up her hands and stopped everything. She read on forums about people who haven't let water touch their face in years. Feeling this was extreme—possibly even pathologic in itself—she kept showering, but without soap or shampoo. The only thing that touched her face was a microfiber cloth and the smallest possible amounts of water.

"It got much worse before it got better," she told me, grimacing. But after two wearying months of oiliness, things started clearing up. There were no longer drastic swings from dry to oily. Her skin remained in more of a constant, steady state. This is a common observation among people who start using less soap. Though there isn't convincing evidence that sebaceous glands actually compensate by secreting more oil when skin is dried out by soaps and astringents, those products do affect microbial populations. The constant washing away of oil-consuming bacterial populations would mean the skin is likely to appear oilier.

The lesson in stories like Dusenbery's is that prescribing antibiotics and steroids often amounts to a well-intentioned shortcut. These drugs can be hazardous to the microbes we need. This shotgun, shut-it-down approach may soon be considered among the relics of medical history, as archaic as miasma theory.

"I don't want to say my skin is great," she told me, "but I get maybe a couple small zits a month."

Now there is a believable marketing slogan.

V

DETOXIFY

"You really need to consult an attorney," my girlfriend kept telling me.

"It's not a crime," became my refrain. "I'm very confident it's not a crime."

It turns out to be remarkably easy, from a bureaucratic perspective, to start your own skin care business. Having encountered so many soap sellers and skin care entrepreneurs who seemed to have dived into this business without any prior knowledge or experience, and knowing that some of them were going to become millionaires if not billionaires, I had to see if it could really be that easy to put a skin care product on the market. I didn't intend to sell anything. I just wanted to learn about the process firsthand.

The plan was to create one skin care product to test out what it was like to actually bring something to market. Not that I was planning to do anything illegal, but if I were to test the boundaries of good conscience, would anyone speak up? Would the government stop me?

Taking a note from my tour through the industry, I knew I

needed a catchy brand and a target demographic: The tagline for my company would be Brunson + Sterling: Menscare for Fucking Perfect Skin. The names don't mean anything; they just sounded right.

I contacted my colleague Katie, an illustrator and designer, to create a logo for the company. We met for lunch at a fast-casual salad place in DC, which turned into a two-hour strategic bonanza, after which we had a spreadsheet that included costs of bulk packaging, a customized website, and Instagram advertising assets. The goal was to combine a minimalist aesthetic with extreme machismo and as many buzzwords and "hot" ingredients as possible.

What *could* we technically put into this . . . product? Could we say it was "natural"? Organic? Healing? Age-defying? Age-reducing? Age-sucking?

The answer is yes, to all of these things. While I couldn't claim that the product could cure specific diseases, just about anything else is fair game. I notified the government agency that oversees the industry, the Food and Drug Administration, that I was going to be selling *a product*, and I gave my address, which is all that's required of new vendors. I didn't have to say what's in the product, or provide any evidence that it was safe, or that it had any effects at all.

Next I turned to the recipe. Many if not most skin care products can be made from ingredients available at any pharmacy or grocery store, so that's where I started.

I went to Whole Foods and bought an array of trendy ingredients: jojoba oil, vitamin C, collagen, acacia fiber (a prebiotic), turmeric, shea butter, honey, coconut oil. I mixed them in a bowl in my kitchen and poured the mix into some two-ounce brown glass jars I ordered from Amazon, printed labels on stickers, and posted the product on a Squarespace website. The process took an afternoon and cost around $150. Brunson + Sterling's flagship product, Gentleman's Cream, was born.

I decided to make no claims about what it does, only list the ingredients and use a vaguely masculine color scheme. Casual elegance, mindful minimalism.

I also decided not to try it on myself or anyone I knew. If we did end up making some sales, plausible deniability could be important. I had no reason to believe anything I was selling was dangerous. In isolation, all of these ingredients are what the FDA considers "generally recognized as safe." But if I did test the product and found any hint of evidence that Brunson + Sterling Gentleman's Cream had zero effect, or was harmful, ethically I'd have to abandon my project. If I found it *did* work—that it increased collagen production, for example, and so truly had "antiaging" effects—I would have the additional guilt of knowing that I was selling a product that messed with people's genes. I wouldn't be able to mention the gene thing on the label without registering the cream as a drug—meaning putting it through all kinds of testing for safety, and then requiring buyers to get a prescription.

I listed a two-ounce jar of my Gentleman's Cream for $200.

• • •

Skin care products are regulated in one (or more) of three categories: soaps, cosmetics, and drugs. These distinctions are more than just bureaucratic line-drawing. They define how these products are regulated, how they are marketed, and how we use them on our bodies.

First, there are soaps. Not every product marketed as soap meets the FDA's definition. The FDA interprets the term "soap" to apply only when a product's cleaning properties come from the combination of a fat and an alkali (as opposed to a synthetic detergent), and the product is labeled, sold, and represented solely as soap. These products are regulated by something called the Consumer Product Safety Commission, along with various other household items such as toys and tools. This commission requires manufacturers to comply with safety standards, but lacks the capacity to inspect each one of the millions of consumer products before they reach the market. Instead it largely reviews products retroactively—after a dangerous problem has occurred. For example, in October 2018, the commission asked Walmart to recall all Ozark Trail camp axes after the commission received reports from consumers that "the axe head can detach from the handle, posing an injury hazard."

The commission's stated goal is "protecting the public from unreasonable risks of injury or death associated with the use of the thousands of types of consumer products." This is the sort of regulation that is touted by many conservative politicians as bad for business, even though consumer-product-related injuries, deaths, and property damage are estimated to cost the nation around $1 trillion every year. And if axes do not need to go through an approval process before going to market, why should soaps?

Of course, soaps are an increasingly small share of the skin care market. Personal care products containing detergents (though they often still say "soap" on the label) are considered cosmetics—which are, like food and drugs, overseen by the FDA.

The Federal Food, Drug, and Cosmetic Act defines cosmetics by their intended use as "articles intended to be rubbed, poured, sprinkled, or sprayed on, introduced into, or otherwise applied to the human body . . . for cleansing, beautifying, promoting attractiveness, or altering the appearance." This includes moisturizers, perfumes, fingernail polish, makeup, shampoo, permanent waves, hair colors, and deodorant.

Drugs, by contrast, are "articles intended for use in the diagnosis, cure, mitigation, treatment, or prevention of disease" and "articles (other than food) intended to affect the structure or any function of the body of man or other animals." Products claiming to "restore hair growth," "reduce cellulite," "treat varicose veins," or "regenerate cells" are supposed to be regulated as drugs.

The *intended* use, of course, varies. It refers to the intention conveyed to the consumer through labeling and advertising claims. So although I may intend to use YouTube hypnotherapy videos to cure my diseased leg, duct tape to treat a plantar wart, or rat poison to relieve a stomachache, my odd intent doesn't make these things *drugs*.

A drug can also be defined by "consumer perception" of its uses. This is why cannabis is a drug, for example, even when sold in the form of a cookie that simply has a picture of a bud on it or something, and doesn't promise to get you "high" or "nicely toasted" or even invoke the word "groovy." The public perception is already there.

Most skin care companies get into trouble only when they are selling a product that makes drug claims but is not registered as a drug. It's up to the seller which path they want to take. The claims made by new skin care products are veering ever more in the direction of drugs. As beauty standards have moved toward a more "natural" look—as opposed to looking made up—more products are promising to change the structure and function of the skin to make it look better, or at least different.

These constitute a growing category of products that could be considered cosmetics *and* drugs. The FDA gives the examples of antidandruff shampoo and moisturizers that claim to provide UV protection. Another is "essential oils," which are considered cosmetics when marketed as fragrances, but drugs if marketed with certain "aromatherapy" claims, such as assertions that the scent will help the consumer sleep or quit smoking.

Though the line between cosmetics and drugs is blurring, the difference in regulation is vast. Before drugs can be taken to market, they require years of clinical trials that cost millions of dollars to accumulate evidence that the product is safe and effective. Cosmetics require no approval or proof of safety.

This discrepancy comes to national attention occasionally. In 2017, for example, major news outlets reported that hairstylist Chaz Dean's popular "cleansing conditioner" product, WEN—which is marketed as extra gentle, with "no harsh chemicals"—allegedly caused a young child named Eliana Lawrence to lose her hair. Photos of the girl spread on social media and came to the attention of Senators Dianne Feinstein and Susan Collins, who met with Eliana. She reportedly told them how scared she was when her hair started falling out, and that she was still teased at school for lingering bald spots.

The FDA had begun investigating WEN in 2014—but only after the agency received 127 customer reports of adverse reactions. By 2016, the number was up to 1,386. The agency found that the manufacturer had itself also received 21,000 complaints of hair loss or scalp irritation, which it had not passed along to the FDA. There is no requirement that it do so.

After all of this, in the wake of Eliana's story making news, the company flatly denied that its product was harmful. A spokesperson said at the time: "There is no credible evidence to support the false and misleading claim that WEN products cause hair loss." The conditioner remains on the market.

Proving with certainty that any particular product is dangerous can be extremely difficult. Unless an illness or reaction appears quickly and reliably in multiple users of a product, things can usually be dismissed as coincidental. Combine that with the laxity of the laws and the meager staffing of the FDA, and it's very rare that products are the subject of regulatory action. It's so rare that when a product is proven to be harmful, it often makes national news.

The novelty of such cases gives many consumers the sense that harm from personal care products is also rare—isolated to the occasional bad egg that swiftly gets pulled from shelves. But even in provable cases, or when a company admits to error and agrees to take a product off the market, the process can take years. In 2017, for example, the youth accessories store Claire's recalled some makeup products marketed to adolescent girls (including "bedazzled rainbow heart makeup set and metallic hot pink glitter makeup") after reports that they contained asbestos, sharp fibers that can cause fatal cancer if inhaled. After the bad press, Claire's chose to recall the products—even though legally it didn't have to. The FDA can't force a company to recall a product. The safety system is an honor code.

It wasn't until March 2019 that FDA commissioner Scott Gottlieb said the FDA had run tests and confirmed the presence of asbestos in the makeup. He used it as an opportunity to remind the public that "the cosmetics industry is undergoing rapid expansion and innovation"—citing $88.2 billion in sales in 2018, up from $73.3 billion five years earlier—and yet "at the same time, the provisions in the Federal Food, Drug, and Cosmetic Act . . . have not been updated since it was first enacted in 1938."

Until the early 1900s, drugs were lumped in with cosmetics and soaps and anything else you might find at the general store. As far as the government was concerned, self-care products fell under the umbrella of "patent medicines." Nothing required a prescription from a doctor. Tonics and elixirs containing powerful narcotics often bore no label at all. If they did, there was no guarantee that the list of ingredients was accurate.

In 1906, Theodore Roosevelt put an end to the party when he signed the Pure Food and Drug Act. The law forbid the manufacture, sale, or transportation of "poisonous or deleterious foods, drugs, medicines, and liquors" in interstate commerce. It also outlawed "misbranded" and "adulterated" products.

More importantly, the law started to define *drugs*. It listed ten active ingredients—including cocaine, cannabis, opium, and heroin—that manufacturers of patent medicines would be required to disclose to consumers. These ingredients were still legal, but they had to be listed on the label. No one should be buying heroin, Roosevelt apparently reasoned, without knowing that they're taking heroin.

Once a product is known to be dangerous or addictive, the question arises: Should companies be selling it at all? The Pure Food and Drug Act was simply a law of transparency. But it opened the door to efforts to outlaw certain drugs that were not safe, followed by efforts to outlaw those that were safe but not effective.

At first administered by the research-oriented Bureau of Chemistry in the Department of Agriculture, these standards were not straightforward. The safety of most drugs depends entirely on how much you take. So in 1927, to deal with the increasingly expansive demands of these questions, the Bureau of Chemistry was made a purely regulatory agency and renamed the Food, Drug, and Insecticide Administration (the third element was dropped from the name three years later). In 1938, the Pure Food and Drug Act was replaced with the more comprehensive Federal Food, Drug, and Cosmetic Act, signed by Roosevelt's cousin Franklin.

Then everything stopped. This law remains the basis for federal regulation of all food, drugs, "biological products," cosmetics, and medical devices. Congress has never updated it.

In the pharmaceutical industry, by contrast, a company cannot bring a product to market until it has undergone clinical trials that have shown it has some benefit and have not shown evidence that it is harmful. The entire process takes years and costs millions of dollars. Even after a pharmaceutical is brought to market, it can't be advertised without a laundry list of adverse side effects—basically the entire second half of any pharmaceutical TV commercial. The advertising is still an ethically dubious proposition, and the clinical trial process is far from perfect, but at least some attempt at regulation and quality control is made. And yet the pharmaceutical industry's offerings are far more distrusted by the public than the skin care products we apply daily to our body's largest, extremely porous organ.

"Right now, when it comes to cosmetics, companies and individuals who market these products in the U.S. hold the responsibility for the safety and labeling of their products," Gottlieb said in a press statement delivered by Twitter thread (a post-1938 innovation). "This means that ultimately a cosmetic manufacturer can decide if they'd like to test their product for safety and register it with the FDA. To be clear, there are currently no legal requirements for any cosmetic manufacturer marketing products to American consumers to test their products for safety."

He ended with some extremely gentle suggestions about what could be done to "work with stakeholders" to "shift the current paradigm." This "could include: mandatory registration and listing, good manufacturing practice regulations, mandatory reporting of adverse events, access to records, mandatory recall, labeling of known cosmetic allergens, and ingredient review."

I asked him in reply (publicly, on Twitter) if he was saying it *should* include these things. At the time he was months away from resigning—a plan he had already announced. "I ask because most people assume these things are already in place—like that FDA can make companies recall products after they've proven dangerous," I wrote. "You'd be on real solid footing as a doctor and head of a regulatory agency saying that should definitely happen."

He didn't reply or clarify. If the head of the nation's regulatory agency does not even publicly say that his agency should have the authority to know what ingredients are in products being sold, we are vastly far from the sort of oversight that would guarantee any sense of safety. Not only can the FDA not force recalls of such products, it lacks the authority even to review the ingredients in personal care products (with the exception of color additives) to determine whether they are safe. As a result, in the U.S., where economic growth has long taken priority over consumer safety, only eleven substances are banned or restricted from use in personal care products. Meanwhile, the European Union and Canada have been reviewing ingredients in personal care products for decades. More than 1,500 chemicals are banned or restricted from these products in the European Union, and some 800 are banned or restricted in Canada. California state lawmakers proposed a bill in 2019 that would ban the inclusion of lead, formaldehyde, mercury, asbestos, and many other potentially harmful compounds from personal care products, which, if enacted, would be the first legislation of its kind in the United States. As of this writing, the effort has not yet been successful.

After learning about the history and present of skin care regulation, I was less concerned about Brunson + Sterling getting me into trouble with regulators.

It remains on the internet today, though I didn't invest enough in advertising to actually induce anyone to buy one of the $200 jars. Ultimately it just felt too evil. Maybe someday I will be able to work up to it.

In the meantime, I am willing to sell the brand for $100 million.

• • •

In an old factory building in the gentrifying industrial neighborhood of Gowanus, Brooklyn, I round a corner and am hit by the smell of lavender. It's coming from down a long hall, behind a door that I buzz, and Rachel Winard answers in a chef's smock. I'm here to make deodorant.

Winard is the proprietor of Soapwalla, a small line of gender-inclusive, conscientiously marketed, minimalist skin products. It's 9:00 a.m., but she's been at it for a while, in a space the size of a large studio apartment that serves as the company's test kitchen, production facility, and distribution center. Four employees busy themselves with various tasks as Winard and I head past the punching bag hanging by the door (she boxes) and into the industrial kitchen, where a large mixing bowl full of off-white powder sits on a counter like a cake mix about to become batter. She acknowledges the cheesiness of the ritual before blessing the powder in gratitude for the opportunity to share it with the world. As we talk, she adds water, stirs, and ladles the mixture into two-ounce jars. I screw the lids on and place the jars in the fridge so the mixture can harden.

The recipe for Soapwalla's eponymous deodorant is top secret. Winard tells me this with a smile, but she means what she says. It's the reason she prepared the ingredients before I arrived. The deodorant achieved a sort of viral fame around 2011, which Winard traces possibly to some public accolades from the actress Olivia Wilde. This was in the dark ages before influencers, and Winard certainly didn't pay anyone to recommend it, or invest in publicity at all. At the time she was making the deodorant out of her own kitchen.

The deodorant's popularity spread, from blog to blog and person to person, in the truest way a product can be recommended: because it is good. Soapwalla deodorant is a cream that has to be applied by finger. It falls in the domain of "natural" deodorants, a technically undefined class that tends to be a signifier of mildness, or of ingredient lists that do not involve molecular names. Natural deodorants also tend not to include the traditional antibiotic compounds long used in stick deodorants, and instead employ essential oils that can mitigate odor, in part by smelling good and in part because they have antimicrobial properties. Instead of the aluminum used in traditional antiperspirants to impede our glandular functions, natural deodorants may rely on blends of clay or other powdered substances that may absorb sebum.

Despite the simplicity and basic commonalities among natural deodorants, Winard seems to have hit on a standout product. For many people who have felt adrift in disastrous experimentation with natural deodorants, Soapwalla shines like a long-awaited lighthouse at the port of freshness. I used it while transitioning off a typical stick, and it worked just as well. But what really sets the product apart to me is the way it's sold. The package is inconspicuous, and the marketing almost nonexistent. Soapwalla does have an Instagram account, but it operates totally outside influencer culture. In fact it does not feature any humans, ever, out of concern that this would create some idealized conception of who the products are for and what bodies should look like.

Winard's entrance into the skin care industry was an unlikely one. At age twelve she was a professional concert violinist traveling the country. She graduated from high school at sixteen and left home on the West Coast to attend Juilliard. While she loved performing, she couldn't countenance the commercial side of making a living in music. Just as decisively as she came in to music, she got out. She took the LSAT and went to law school at Columbia.

Her second day of law school was September 11, 2001.

In the weeks and months after, she would volunteer at Ground Zero. As with many others who helped in the response, inhaling the remains of all that had become dust, her health took a turn for the worse. It was around this time that, as she put it, her body "started attacking itself." Whether it was due to some exposure there, or the emotion of it, or totally coincidental, she can't be sure. But within weeks she went from a picture of health to being barely able to get out of bed. She was completely sapped, her energy gone. It didn't feel like depression, she says, but like life had simply been sucked out of her.

It started with her skin.

"I didn't have acne as a teenager—never had any dry or oily skin issues," she told me. But a blotchy red rash that appeared on her face and arms that fall progressed to joint pains and fevers. It took a year of seeing various doctors before she was diagnosed with lupus—an autoimmune disorder notorious for its varied manifestations.

"I think the skin can be sort of the canary in the coal mine," she said. "Like, when you have more systemic issues happening, you'll see it on your skin before you necessarily feel it—or understand that what you're feeling is something that should maybe be acknowledged."

She gently corrects my lid-screwing technique.

Winard went on all the standard immune-suppressing drugs to treat the lupus, and the symptoms sometimes got a little better, but her skin just kept getting worse—redder and itchier, burning and painful. "When it was at its worst, I couldn't let water on my skin," she recalls. "And so I was that desperate consumer who would scour the shelves, looking for anything that said it was hypoallergenic or for sensitive skin, natural, organic—all of this terminology that was just starting to be used then, in 2003."

Still, as she tried to get cleaner and cleaner, to rid herself of whatever was causing this, it only seemed to get worse. Until one night, out of "utter desperation, when I couldn't sleep because I wanted to rip all my skin off, I thought, 'Well, I'm going to have to make something because I can't live like this.'"

She started mixing and iterating, attempting to find something mild enough that she wouldn't smell bad, but would give her skin a break. In the process she hit upon the formulation of her now beloved deodorant and started using it. During this period of self-exploration, she also took a sabbatical from law and went to India for a year to "reset." She started doing yoga and became more conscious of what she ate.

Somewhere in the mix of it all, she says, her skin cleared up and her health returned. It was as if her immune response went back into normal gear. She doesn't attempt to distill the explanation to any single change—to stopping the ultra-aggressive washing, starting to apply her minimalist concoctions, or to the existential catharsis of leaving New York. She suggests the answer involved all of the above, and more. Disorders of the immune system tend to be inseparable from stress, sleep, physical activity, and the overall mix of what we put in and on ourselves.

It's an experience familiar to many people with chronic diseases who go through periods of remission and health, sometimes for no apparent reason. The good periods become a sort of North Star. After serious suffering, moments of relief can make doing whatever you're doing at the time seem like it's the answer. No doctor's advice will be more compelling than the instinct to keep doing what you're doing. When she came home, Winard tried to hang on to as much of her new lifestyle as possible. And for the most part, it has worked well.

Having finally found a formula for deodorant that worked for her, Winard decided in 2009 to start selling it to others who might be in similar situations. She didn't get any venture capital or even advertise her product. But word spread among friends locally, and then into the blog-addled internet of 2010. Within two years she'd go from filling an occasional order in her spare time to leaving her law firm to run the company.

Exactly how or why this deodorant works especially well for many people—and, for others I spoke to, doesn't—isn't totally clear. Winard includes clay to absorb moisture, as has been done for centuries, around the world. Microbiologists I spoke with suggested that there may be something to particular combinations of powders and essential oils that balance or shift microbial populations away from the odor-producing species while still allowing other species to thrive. This would mean that seeing effects could take time.

After I had sufficiently convinced myself that it was possible to quit deodorant altogether, I went back to using Soapwalla every few days. It gives me the certainty of smelling unobjectionable, the odds of which are always somewhere below 100 percent when I don't use anything at all. That experience of personal trial and error was my high-states induction into the self-experimentation side of skin care. When a product works, there's truly no going back.

But more important than creating a beloved armpit cream, Winard's most salient contribution to the skin landscape might be her way of existing in an industry where there are so many ways to go astray. She seems to be a kind of proof that it's possible to exist in the beauty or wellness industry—and others—without selling people on an idealized standard of what they're supposed to look or smell or feel like.

There are other models for entrepreneurship that don't involve encouraging people to apply more and more products, or to look a certain way, or to live in constant pursuit of an unobtainable standard. The minimalist movement in skin care is gaining momentum.

Adina Grigore is another breakout star from the New York indie scene—and an especially unique case because she both gave up showering and started a skin care company. It's called S.W. Basics, and the philosophy of the company is that most people should be doing way less to their skin.

"So what I'm trying to do with the line is go, 'Leave your skin the fuck alone,'" she tells me. "Just leave it alone as much as you can."

Grigore is in her thirties and newly left New York for Denver, where she now leads the small company. The S.W. Basics line seems to be the closest any mass-market company comes now to selling skin products under the pretense of doing almost nothing to our skin.

The bestselling products are a rosewater spray (called simply "Rosewater") and a cream for dry skin ("Cream"). Grigore recounted to me in an impassioned monologue the story of how her own health inspired her business: problems with acne led to challenges to identity and issues with control.

"I was just covered in, essentially, a full body rash," she tells me. She was told she had folliculitis and started using a steroid cream all over her body. She stayed on it for two straight years. Usually steroids aren't recommended for more than a few weeks at a time because, while they're effective at shutting down the immune system, that is not an inconsequential act. In the long term, they cause the skin to break down. She describes her skin becoming visibly thinner.

This is where Grigore, like Maya Dusenbery, took matters into her own hands. "I can no longer handle waking up with bloody sheets because I'm scratching my skin in my sleep. Out of desperation I was like, that's it. I've spent all this money and done exactly what they told me. Now I'm ditching literally everything—I'm not putting anything on my skin."

Within a matter of days, she recalls, "everything was better."

Now she keeps alive the spirit of skin care in the minimalist style—for people who like the fragrances and the sensations and the ritual but pretty much want to leave their skin alone. She is open about the fact that most of what she sells could be easily made by anyone in their own kitchen.

Yet despite and possibly because of her candor, Grigore's concept has had tremendous success. She started out with money from angel investors and recently signed distribution deals with Target and Whole Foods. Even with all the products these places carry, to be included in their stores is an elite and highly coveted achievement. Dermatologists also sell her products directly to their patients, taking a percentage of the sale as what would technically be a commission. (It is considered unethical for doctors to take a cut of each prescription drug they prescribe, in part because having a financial incentive to prescribe certain drugs could bias the scientific judgment of the practitioner. But doing so with skin care products raises no such flags.)

All of this ideal placement also obviated the need for a strong sales pitch from Grigore—of the sort you see when companies are still trying to get noticed on Instagram. She can continue speaking her truth: that less can be more. As she put it to me, "The things that you think might work for you, because everyone else says they do, just might not. People aren't patient enough with themselves and with their bodies, and then they're actually being told by literally everyone, 'No, you don't need the patience, I'm the thing that'll fix you overnight.'"

Once she stripped down her routine, Grigore became especially attuned to the other things in life that affected her skin—food and sleep and stress. These effects become easier to appreciate when there are fewer variables being applied. As she describes the experience, cutting back on products put her in touch with this less-talked-about side of skin care: caring for everything inside of one's skin. What might be called self-care, or simply health.

• • •

If it seems redundant or odd that so many people start their own lines of skin care products, the instinct to do so may trace to the deep-rooted distrust of the current marketplace. While there can be many honest and well-intentioned sellers, it does not take many unscrupulous actors to ruin consumer confidence. In an attempt to help restore that trust, Senators Susan Collins and Dianne Feinstein introduced the bipartisan Personal Care Products Safety Act to Congress in 2017. They argued at the time that "it makes no sense for each company in a multibillion dollar industry to have to make its own determinations about minimum safety standards."

Writing in the medical journal *JAMA Internal Medicine*, the senators warned: "There is no other class of products so widely used in the United States with so little regulation." They concluded, in no uncertain terms, "The lack of oversight is a broad threat to public health."

The bill would simply require companies to disclose *what* is in the millions of products that we slather over ourselves every day—not to prove safety, just to register the product and say what's in it. When consumers report serious adverse effects from a product, the companies would be required to report that to the FDA. If the agency sees a problematic pattern, it would have the authority to require warning labels and, if a product is causing serious problems, to recall it.

The bill would also establish an independent review process for ingredients used in personal care products and authorize the FDA to look at all available information on particular chemicals to determine whether they are safe. The FDA would be required to review at least five (*five*) chemicals or categories of chemicals per year, chosen based on input from consumers, medical professionals, scientists, and companies.

It's surprising to most people I talk to that none of this regulation is in place already, especially in an industry that's ostensibly all about purity. Can any consumer really be operating autonomously when their access to information is incomplete, and the playing field skewed so heavily toward sellers?

For decades, industries have successfully convinced the public and its lawmakers that regulation would raise the price of products and would be bad for jobs. Requirements for testing products would increase the cost of basic goods, because corporations would pass those costs on to consumers. This would amount to an archaic and even dangerous tax on soaps, which society deems vital to public health.

At the same time, others worry that regulation would also raise the barriers to entry and keep new competitors from breaking in. This is what many people love about skin care: a feeling of meritocracy and low barriers to entry. Small companies produce small products, and what's good should genuinely rise to the top, by word of mouth, because something actually works. Skin care is the forefront of a widespread democratization of health—a shift away from centralized medical authority. Legally, anyone can enter. The gatekeepers are not gone, but their gates are much lower. No one needs health insurance or a prescription to partake. The sellers do not need to go through training or assume hundreds of thousands of dollars in student loans. They do not even need to have much overhead. Companies can be run out of one's apartment. Marketing can be done on Instagram.

Many consumers are beyond ready for this shift of power. Unlike in many other medical specialties, dermatology patients can often see whether their treatment is working. A cardiologist may prescribe a blood pressure medication or a cholesterol drug that is supposed to decrease a person's chance of dying decades later, but it doesn't change how they look or feel. Likewise, only an oncologist can assess just how well chemotherapy is eradicating a cancer. But anyone who looks in the mirror has an instant set of meaningful data points about the state of their own skin.

Maya Dusenbery, whose personal health journey was so similar to Winard's, thinks the role of doctors and scientists in all this should not be to try to regain their spot as the primary keepers of knowledge, or to position themselves as the sole arbiters of truth. Instead, it is time to break down the old dichotomy in medicine between "mainstream" and "alternative." I had come to think of this in the clichéd way: what science has proven, and everything else. But it is more complex than that. Instead, experts and authoritative agencies might help break things down into four categories: what clearly works, what plausibly *might* work but hasn't been studied, what's totally implausible, and what's proven to be useless or harmful.

Dusenbery felt well equipped to take the path of radical minimalism and experimentation with novel products because she is a journalist who has covered science and medicine for years. She wrote a book of ethnographic history about the medical profession's systematic biases against women. She has a unique sense of the problems with paternalism as well as the problems with anarchy.

"There are definitely aspects of the communities that have formed around sharing health and beauty knowledge that are bringing people, especially women, together to take control of what's long been a male-dominated field," she says.

Indeed, the internet is loaded with forums and accounts of skin gurus who are clearly sparking debate and building followings. For example, the popular podcast Forever35 maintains a private Facebook page where listeners go to post "about self-care and wellness." The community was over 17,000 strong when I last checked, and one of the most popular tags is "skin care," with discussions of vitamin C and acne and things familiar to any skin care space. The discussion manages to be lighthearted and affirming, bringing into public space the daily routines that for so long were undertaken in private and discussed only with close friends. In this way the social currency of skin care is no longer all about the outcome—how you ultimately look—but about the process.

The trick is keeping perspective on the overall costs and benefits, and not letting other people determine your value system—insofar as that's possible. Several years after her product cleanse, Dusenbery only uses products that she feels are actively adding value to her life, and out of curiosity and for pleasure more than from any feeling of necessity. She spritzes rosewater on herself, and she dabbled in snail serums. In winter she tried beef tallow as a moisturizer, as well as a bear tallow lip balm made by a dreadlocked herbalist in eastern Oregon from a bear she shot herself. "I use makeup sometimes, when I feel like it, but not in a way that I feel like I need to—nothing that I have to do every day," Dusenbery says. (She does feel she couldn't have done this back when her acne was at its most severe. Given the way acne is imbued with judgment about poor hygiene, the professional and social penalties would've been too intense. "We just don't live in a society where that's really possible.")

Many people want to do less, be more minimalistic and "natural," but still have the grounding ritual, time for oneself, social signifiers, and social bonding that partaking in cleaning routines provides. In a historic context, the communities that spring up around skin care products and routines and beliefs are more true to the essence of being *clean* than any of the products themselves.

The solidarity and passion that accrue behind product lines and brands are sometimes also obtainable in their absence. Abstainers bond over having given one or many things up. Environmentalists and "no-poo" (referring to shampoo) devotees find identity in the act of going without. I found that once the subject is on the table, a lot of people are actually pretty eager to talk about their hygiene beliefs and practices. The subject has the effect of instantly breaking down a barrier, like sharing a secret that almost no one else knows, when actually all they've told me is how often they shower. I'm not suggesting that as an opening question to a stranger. But breaking down conversational taboos is actually an important step in challenging the standards that beg to be challenged. Once you start hearing about all the things people do and don't do, use and don't use, couldn't live with or without, standards of normalcy fade. Then you can focus on what actually matters to you.

VI

MINIMIZE

Living among the rolling hills of corn in Pennsylvania and the supine expanses of corn in Indiana are people who almost never get asthma and have very few allergies. And they have, by all credible accounts, abnormally good skin.

The traffic had been humming along on the two-lane Indiana interstate when, in the middle of a sunny Sunday afternoon, it suddenly slows to a crawl. There is a horse-drawn buggy on the road, decorated in back with a three-foot reflective triangle. This happens a lot in Amish country. The buggies sneak up on you fast when you're going 70 miles per hour and they're going 10. After a spate of highway collisions, some buggies have now been outfitted with lights—breaking with the community's traditional avoidance of modern technology, but preventing catastrophic death.

The slowed traffic affords an opportunity to peruse the roadside stands selling furniture, quilts, and candy, and the people in nineteenth-century attire working in the fields beyond. Some of the white wooden homes I pass have phones hooked up in the yard—a way of connecting to the outside world without letting it become *too* accessible.

Mark Holbreich, an allergist and immunologist who has worked in Indiana for three decades, noticed something different about the Amish that went beyond their low-tech lifestyle. He does research affiliated with Indiana University, where I went to med school. Our hospitals and clinics were close enough to the Amish populations in the northern part of the state that we saw a not-insignificant number of Amish patients.

"My first observation is that their skin is particularly clear and healthy-looking," he tells me. He also noticed that the Amish communities he served seemed to have low rates of asthma and allergies—that those patients who came to him thinking they had an allergy actually didn't. "We rarely see eczema or skin problems," he says.

He wondered if this low incidence of skin conditions was somehow related to the genetics of the Amish—who emigrated from Switzerland two centuries ago, and famously keep a tight gene pool—or if it had more to do with their lifestyle. Holbreich dug through the research and uncovered some European studies indicating that children who grew up on farms tended to have lower rates of asthma and allergies compared to city or suburban kids.

In 2007, Erika von Mutius at the children's hospital of the University of Munich reviewed fifteen studies of immune system function conducted over the prior decade in rural areas in Europe (Switzerland, Germany, Austria, France, Sweden, Denmark, Finland, and Britain). Almost all the studies she reviewed found that farming communities had very low rates of hay fever and allergies. Several of the studies found less asthma and sensitization to allergens among "farm children" versus "nonfarm children." (These aren't skeptical quotes—I just like the terms.) In the journal *Proceedings of the American Thoracic Society*, von Mutius deemed this the discovery of a "farming effect" on the immune system.

When Holbreich tested over 100 Amish children in his Indiana clinic, he found rates of asthma and allergies that were not just low by U.S. standards. They were even lower than rates in Switzerland—just 5 percent, compared to 7 percent in Swiss farm children, and 11 percent in Swiss nonfarm children. He couldn't exactly say why, but his hypothesis followed the European researchers', involving "some impact of early life exposure to microbes which we think are inhaled, swallowed, and on the skin."

To test this hypothesis, Holbreich teamed up with a group of researchers, including von Mutius, to compare the allergy rates of two genetically similar farming communities, the Amish in Indiana and the Hutterites in South Dakota. Both originated in the same region in Europe during the Protestant Reformation and came to America between the 1700s and 1800s. Both have since remained relatively isolated, with lifestyles that are similar in many respects—especially those that might influence the immune system. (They don't keep many indoor pets, generally have large families living together eating similar diets, are exposed to low rates of smoking and air pollution, and have comparably high rates of breast-feeding.)

In August 2016, Holbreich, von Mutius, and their colleagues rocked the world of immunology when they published their findings in the eminent *New England Journal of Medicine*. Despite all the similarities between the two groups, rates of asthma were four times lower and rates of allergies were six times lower among the Amish children compared to the Hutterites.

The key difference between the two groups, the researchers

determined, was the proximity of their homes to their farms. Amish kids grow up extensively interacting with the farm environment: the animals, the soil, and the airborne sediment and microbes that a farmer inhales. The exposure starts in infancy, as parents strap their babies to them while they do the rounds on the farm.

Hutterite kids have a different experience, removed from such direct contact with farming life. They live on large communal properties with homes arranged around a central farm. The men are driven out each morning to work, but children are not allowed to accompany them. The Hutterites have also embraced modern agricultural technology, which means that much of their work is more highly mechanized than the Amish hands-in-the-dirt approach.

"The Amish and Hutterites are very hygienic," Holbreich tells me, careful to make the distinction that they do not see high rates of preventable infectious diseases in either group. That is despite—and, he believes, because of—their exposure to lots of microbes. In some bacteria, for example, proteins known as endotoxins stimulate the immune system. The researchers found that levels of endotoxin in the dust of Amish houses were seven times higher than in the Hutterite homes. They also looked at the children's immune systems and found that the numbers and types of immune cells showed "profound differences."

As if this weren't enough, the scientists used "electrostatic dust collectors" (vacuums) to collect house dust from both populations and puffed it into the noses of mice. The mice exposed to Amish dust, compared to those exposed to placebo dust, developed less-reactive airways and lower levels of allergy-related cells.

Holbreich says his grandmother always said to "eat some dirt every day and you'll be healthier, or something like that." He didn't, and he doesn't recommend letting children roam the wilderness in feral packs or, as I suggested, packaging the dust from Amish farms as a miracle allergy cure. "The science is just not there," he says.

What the science does suggest is a far more complicated relationship between our bodies and microbes than we previously understood.

• • •

When there is inflammation in one part of the body, the rest of the body knows. Messages travel through intricate channels of white fluid coursing throughout our bodies, connecting our heart and our skin. Just as the circulatory system deals in blood, the lymphatic system deals in lymph, the liquid carrier of immune cells. As crucial as it is to our existence, the whole system was discovered only very recently.

In 1622, the Italian scientist Gaspare Aselli was dissecting a dog and found "milky veins" that looked like they contained white blood. But he did not know what he had found. *What was this, a demon dog? A second circulatory system? Do all animals have white and red blood? Other colors, too?* His contemporary, the physician William Harvey, proposed that humans also had an entire system of vessels carrying white fluid. It ran in parallel to the blood vessels, and it would be called a "lymphatic system." But he couldn't actually explain it.

It was not until 1962, at a meeting of the New York Academy of Sciences, that a pathologist from Oxford named James Gowans reported discoveries that explained how this fluid created long-term protection from disease. The lymphatic system was separate from the circulatory system, he explained, but immune cells could be passed back and forth. He described experiments in which cells from the lymph of one rat could be injected into the veins of another. When the cells were marked with adenosine so they could be tracked, Gowans watched them rapidly exit through tubes and into pools he referred to as lymph nodes.

The pea-sized organs are intersections of the lymph vessels that run all over our bodies. When you have an infection, the lymph nodes in the area become loaded with white blood cells and swell to several times their usual size. When doctors feel around the base of a patient's jaw during a physical exam, they are searching for swollen lymph nodes. But even when lymph nodes aren't enlarged, billions of white blood cells—lymphocytes—are passing through daily. Gowans explained how removing these lymphocytes from rats left them immunodeficient, unable to mount an inflammatory attack. But the same lymphocytes, when infused into the rats, could entirely restore the ability to fight infections.

Lymphocytes are sometimes in our blood, and sometimes our nodes, but most often they are out in the tissues of the body, essentially doing surveillance. They are looking for antigens, often described as "foreign" material (microbial or otherwise) that gets into or onto our bodies. The lymphocytes then catch a ride in the lymph back to the nodes to drop the hot gossip. If all is clear, they stay quiet and simply go back out. But when they've found something, they gather a furious crowd of other lymphocytes to go out and attack the source of the antigen. The process is referred to as inflammation.

Inflammation can kill us, and it can save our lives. The difference is contingent on constant calibration of the system, so that lymphocytes know when and how aggressively to react to any given exposure. This requires constant training. As a pediatrician might explain it to a child who likes dinosaurs, the immune system can be trained to attack a particular target sort of like velociraptors are trained, by having Chris Pratt dangle a hunk of the target's meat in front of them. This sort of limited exposure, which is also the basic idea behind vaccination, prepares immune cells to identify and fight enemy invaders. The trained immune cells then work like trained raptors, roaming the Jurassic theme park (the body) looking for their targets, and hunting them relentlessly and ruthlessly—but not wantonly killing anything that moves.

This can easily go horribly wrong, as anyone who has worked with raptors knows. Our immune system is similarly powerful and willing to use decisive force. Without proper training (through exposure to the antigens it's meant to target as well as to benign things it's not), our immune system is more likely to attack harmless invaders, and even our own cells. When the predators in the dinosaur park start killing one another, you've got a hit movie on your hands. When the immune system starts attacking the self, you get autoimmune diseases.

These diseases result from a mix of genetic propensity and exposures over the course of a life. Some people are likely to develop an autoimmune condition no matter what they do, but the odds are at least affected—if not largely determined—by exposures that train the immune system. Exposure early in life is the key. By three or four years of age, a child's microbiome is established and the immune system has completed much of its training. Even if a person does not develop an autoimmune condition until later, it seems that the foundation of inflammatory processes is laid in the first few years of life.

In wealthy countries around the world, people now spend more than 90 percent of their lives indoors. Friends and family are not allowed to touch babies unless their hands have been scrubbed or coated in antibacterial gels. The indoor air is lacking in the wealth of bacterial particles that used to temper our immune systems. Our diet is hyperprocessed and cleaned and low in fresh fruits and vegetables—which are naturally loaded with bacteria. An average apple contains 100 million microbes.

In aggregate, these and all the many other precautions we've taken—with the best of intentions—to protect ourselves and our loved ones from disease, and to appear constantly and meticulously "clean," have had at least some role in changing how our immune systems develop.

This idea has been slow to catch on, though its seeds were planted decades ago. In the 1980s, David Strachan, an epidemiologist then working at the London School of Hygiene and Tropical Medicine, started off studying mold exposure as a cause of asthma. He quickly realized the causes were more complex than any single home infestation. Just as some disease is caused by the presence of microbes, other diseases might be caused by microbial absence.

Based on a national survey of British children, Strachan noted that babies born in homes with many siblings were later less susceptible to eczema and to hay fever than kids with few siblings. Some 10 percent of firstborn children had allergies, while those with two older siblings were half as likely to have allergies. The rate dropped by half again among people with four or more older siblings—meaning that firstborn siblings are four times more likely to have allergies compared to fifthborns.

As everyone who has interacted with kids knows, they are walking virus- and bacteria-distribution machines. Assuming that more children in a house meant more germs being shared, Strachan suggested that infections in early childhood protect people against allergic disease.

His "hygiene hypothesis"—as it would become known—spoke to many scientists because it also happened to explain the recent increase of allergic disease in the developed world. Families were getting smaller and more isolated, the rates of childhood infectious disease were dropping, and most people were practicing hygiene and cleanliness at levels never known in human history.

Antibiotics were a revolution in health, and made us able to survive infectious diseases that would've once been a death sentence. They are part of the reason that over the past century, the leading causes of death and disability have shifted heavily toward illnesses like cancers, cardiovascular disease, diabetes, and other metabolic diseases associated with obesity and sedentary lives. At the same time, some chronic conditions seem to be fueled by the fact that so many of us are now not being exposed *enough* to the world.

The basic idea is that as our immune systems encounter fewer benign triggers to teach them how to function, they are attacking our bodies more often than they did in the past. This is offered as a contributing explanation for things like the apparent rise of peanut allergies and gluten intolerance. Today there are places where one in four kids has eczema. Hay fever was once so rare as to be fashionable—a sign of status and affluence. The condition was almost exclusive to isolated elites, while farmers—who were regularly exposed to higher levels of pollen—almost never got it. Since the 1950s, rates of hay fever—as well as multiple sclerosis, Crohn's disease, food allergies, type 1 diabetes, and asthma—have all roughly tripled.

Immune and allergic conditions clearly appear to have increased in tandem with industrialization. Today the prevalence of food allergies in preschool children is as high as 10 percent in Western countries, and is on the rise in rapidly growing countries like China. Type 1 diabetes is far more common in Europe and North America than in the rest of the world, and the percentage of kids with the condition is rising more than 3 percent per year across Europe. Ulcerative colitis and Crohn's disease are twice as common in Western Europe as in Eastern Europe.

To test the effect of isolation and industrialization on these diseases, a landmark study started in 2008 followed children from three countries with close genetic backgrounds but clear differences in rates of allergies and type 1 diabetes: industrialized Finland (which has high rates of both), rapidly modernizing Estonia (where rates have been increasing), and Russia (where both conditions are, comparatively, still rare). Researchers monitored monthly stool samples from over 200 children in their first three years of life. They found that Finnish and Estonian infants have distinct populations of gut microbes compared with Russians, which could account for the difference—not a genetic difference, but a difference in accumulated exposures.

As evidence mounts, though, some experts are worried by any implication that hygiene can be overdone. Sally Bloomfield, for one, is a self-described hygiene advocate. An honorary professor at the London School of Hygiene and Tropical Medicine, where David Strachan developed his theory, she shudders audibly over the phone when I say "hygiene hypothesis." She worries that the term can be easily misinterpreted to mean all hygiene practices are bad, and that we could be propelled back into the preindustrial era of infectious disease outbreaks of the sort that, though less common today, still have the potential to cause catastrophic pandemics.

Like many scientists who study hygiene, Bloomfield has collaborated with the soap industry—she spent seven years at Port Sunlight working for Unilever. Some of Val Curtis's research has also been funded by industry, and she worked with Procter and Gamble, Colgate-Palmolive, and Unilever on hand-washing advocacy campaigns. This I note for purposes of transparency; not to imply that intellectually honest collaborations don't exist. Bloomfield advocates for a measured approach, what she calls *targeted* hygiene—focusing on practices that most affect the spread of disease. For example, she advocates regular hand washing, and also recommends washing hand towels daily, but acknowledges that bathing and showering may not be strictly necessary. She concedes that we are just beginning to figure out what we should expose ourselves to and what we shouldn't—what we should clean off and what we should welcome. The challenge is one of achieving a healthy balance, not of simply doing more or less.

Bloomfield counters my questions about over-cleaning by noting that, on the whole, human life spans are increasing, and people are enjoying more healthy years, generation after generation. Even if we overdid it and some new issues arose, isn't living longer what really matters?

• • •

After leaving Amish country, I head to a secretive, heavily guarded compound just outside Chicago called the Argonne National Laboratory. Argonne is an enormous bureaucratic facility with buildings marked Area 400 and Area 500, started by the federal government in 1942 as an early part of the Manhattan Project.

An armed guard greets me at the gate and asks me my business.

I say that I'm a curious taxpayer, and she doesn't laugh, and instead asks me to turn around and go back to a security office for clearance. My rental car is searched, and finally a gate opens to allow me to enter the mazelike grounds. I keep looking for clues of some government conspiracy that would necessitate this level of security—an errant bag of humanoid organs discarded on the side of the road, a forgotten military-grade hoverboard idling in a field, the distant cackling of a mad scientist.

I would later learn that in addition to the long-secret nuclear work, chemists at Argonne also discovered elements 99 and 100 and first visualized a neutrino. The facility was home to a proton accelerator called the Zero Gradient Synchrotron that allowed physicists to track subatomic particles, and to the first nuclear reactor that could reprocess its own fuel, reducing atomic waste and avoiding additional disasters like Chernobyl and Three Mile Island. Researchers at Argonne still do national security work, including developing defenses against bioterrorism and cyberattacks.

It is also here that Jack Gilbert studies the microbes that populate our skin and bowels. I eventually find my way to a warehouse-like building that houses his labs and office. As I step inside, I hear my name called from the end of a long hallway decorated with scientific illustrations of double helices and viruses. Gilbert is bounding toward me, waving. There is that period of silence that happens when you've greeted someone too early and don't know what to say next, or if you should make eye contact.

He's dressed in jeans, sneakers, and a brown T-shirt, his face unshaven and hair tousled. He does not noticeably smell bad, nor does he look unhealthy, but rather like he doesn't give a damn what

I think of him, because he's worried about more important things, and he has been working on them since the moment he woke up. He is, in other words, a scientist.

Gilbert is a sort of prodigy in the microbiome field. At forty-one, when we met, he was a full professor at the University of Chicago overseeing five affiliated microbiology labs, including the one at Argonne. He was one of Mark Holbreich and Erica von Mutius's collaborators on the Amish allergy study.

"So, I shower," he says, cutting right to the chase. "I do shower, even though I know the potential implications of it. Not every day, and I don't tend to use a lot of soap when I do. Occasionally I will wash my hair with a shampoo, but I use a very, very light shampoo."

Gilbert points out that we are always covered in microbes—even while in the shower. Removing microbes from your skin just opens up space for more microbes—including pathogens—to take refuge. If you think of your skin as a house party (as people do) and you have room for twenty guests, you want to invite at least twenty people you at least sort of like so as to minimize the space for people you don't like—who might end up hanging out late into the night, and then asking for breakfast the next day, and then messing up your bathroom and emptying your fridge and eventually burning down your house.

Even if we lived alone and touched nothing, we would still pick up bacteria. "We're all breathing in microbes," he explains. "When there's smog around, you get very different bacteria and fungi growing on those air particles. Even if those microbes aren't causing diseases like the flu—they usually aren't—they can sensitize the body's immune system."

In Beijing, he explains, if you open your windows during a smog event, the bacteria and the fungi that are growing on airborne particles have the potential to be highly pathogenic, or disease-causing. Breathing in a bunch of new microbes the body hasn't seen before could also send people into autoimmune flares—the sort of haywire overreaction that happens with food allergies. Bacteria and fungi love growing on particulate matter that hangs in the air from burning fossil fuels. Inhaling microbes that ride on air pollution is, Gilbert notes, "not the kind of microbial exposure that your ancestors got used to."

Just before I arrived, Gilbert had been on the phone with Procter and Gamble, whom he has been working with on solutions for improving air quality in homes. But while nobody wants to breathe smog, at the same time, a hyperfiltered existence where there are zero inhaled microbes may not be ideal.

Gilbert essentially argues for balance: "I needed to vaccinate my kids, because I don't want them to die. I need to get them to wash their hands after they go to the toilet, just in case there's a viral or a bacterial disease passing through the population. But do I need to sterilize every surface in my kitchen every time I cook? You can wash it with warm soapy water maybe if you've cut up chicken, but do I need to sterilize it? No."

He explains that if you really wanted to kill all the bacteria on your countertop, you'd have to leave a disinfectant (like Clorox) in contact with the surface for ten minutes. The product isn't "killing 99.9% of germs" in the way that anyone actually uses it—a quick wipe-down.

This was, both in concept and in practice, misguided. And the magnitude of its effects on our lives is now starting to become clear.

• • •

"You know Lifebuoy?" Luis Spitz asks me, almost rhetorically, clearly assuming that I do.

I admit that I don't, and he glares into the back of my eye and waves a finger at my nose.

"You must know Lifebuoy! You are a medical doctor writing a book about soap and do not know Lifebuoy?" he berates. "This is no good."

He leads me urgently through his sprawling basement of soap memorabilia to a row of trolley-car signs. In the late 1800s, the cars had metal slots over the windows that held signs to advertise products (similar to what New York's subways have today, featuring Perfectil and others). The trolley ads for Lifebuoy Health Soap are distinct for their medical promises. In one, a smiling family stands beside the claim, "Families using Lifebuoy have fewer colds and fevers." In a preantibiotic era when fevers were much more often deadly, this was not a claim to scoff at. Another shows a woman saying to a child, "My children must *purify* hands before eating."

More than any other brand, Spitz explains, Lifebuoy sold the idea that soap was essentially medicine. The ad copy—as well as Lifebuoy's life-preserver symbology—brought soap into the world of "health" in a very specific way. Introduced when germ theory was still a novel concept, Lifebuoy, and other soap brands that would follow, did more to promote that theory than any messaging from the scientific community.

Lever Brothers introduced Lifebuoy in Great Britain in 1894. Its active ingredient was carbolic acid, a compound derived from coal tar, which the British doctor Joseph Lister (namesake of Listerine) had discovered could be used as an antiseptic in operating rooms. Carbolic acid also gave Lifebuoy its distinctive red color. The company soon started manufacturing the soap in the United States, where it was marketed as "The Friend of Health" and "A Life-Saver." By 1915, the product was rebranded "Lifebuoy Health Soap." Advertisements for it were even the first to use (as best I can find) the now-ubiquitous term "skin health."

The soap sold well, especially as its rise coincided with the 1918 influenza pandemic that killed some fifty million people. Because of its strong association with health, Lifebuoy remains one of the world's bestselling soaps (although it no longer contains carbolic acid, and it's no longer sold in the U.S.). It is one of Unilever's biggest brands in India, where it is marketed as a defense against germs that cause stomach infections, eye infections, and respiratory infections.

Though the general health and germ-fighting ad copy that launched the brand are still what it's known for today, it was an incident in a locker room in 1926 that would give the brand its most lasting impact. As the story goes, one hot day after a game of golf, the brother of the then-president of Lever Brothers entered the locker room and, as Spitz recounts, "greatly disliked the prevailing odors." This was one of those "aha moments," when the sensory system is overwhelmed by something so horrible that the need for a solution is undeniable.

Lifebuoy's first "Perspiration Odors" ads were soon public. The term would soon broaden to "Body Odors" and then simply "B.O." Spitz explains that Lifebuoy is responsible for popularizing the concept of "B.O."—which was, and remains, an advertising term—and that it brought the brand into the big leagues of soap. Between 1926 and 1930, sales quadrupled.

The term was so compelling, and the implication in the ad such a powerful association, that people who did not have any issue with odor at all bought it as a preventive measure. This was insecurity- and fear-based marketing of the sort that would later come to dominate the industry.

Lifebuoy's term "deodorancy" would become "deodorant," which would become a stand-alone product that people would apply every day in addition to cleaning with soap. The entire conceit was that body odor was caused by bacteria, and so a soap with a bacteria-killing compound was necessary to prevent odor.

This was a marketing concept, not a scientific one, and other entrepreneurs took notice. Watching the rise of Lifebuoy, the Chicago-based meatpacking company Armour decided to get into the deodorant business. Armour had been making soap for a few years because it had excess animal fat. It put hexachlorophene in soap and claimed to have tested its abilities to reduce odors. Dial was introduced in 1948. The name conjured a clock face, and promised to keep users "Fresh Around the Clock." The first advertisement stated "Stops Odor Before It Starts."

Three years after its introduction, according to Spitz, Dial passed Lifebuoy as the most popular antimicrobial soap. Armour bet heavily on advertising and lost $3 million in the first two years, but by 1953 it made $4 million in profit. Enamored of the promise of killing germs, Americans made Dial the bestselling soap in the country.

It was even chosen to be the first "space age soap." Astronaut Alan Shepard carried a bar of Dial on the historic first U.S. manned space flight above Earth in 1961. And then, of course, there was no turning back. We were in space, and we were sterilizing ourselves in the shower.

Shortly after, Procter and Gamble moved into the growing deodorant-soap arena. In 1963, the company launched Safeguard New Deodorant and Antibacterial Soap. It contained an antibiotic called triclocarban. Few thought to question the wisdom of daily application of antibiotics, even when studies came out showing that people who used hexachlorophene soaps and deodorants accumulated the compound on their skin. By the 1970s, other studies found that hexachlorophene could be absorbed *through* the skin into the body, where it affected the nervous system.

In 1972, the FDA issued a recall of consumer products containing more than 0.75 percent hexachlorophene, but the agency had no registry of what products contained the compound, or in what quantities. Disclosure of ingredients to the FDA was voluntary, as it remains today. According to one historian, by the time the FDA acted, about four million pounds of the compound were being manufactured annually for use in medical and cosmetic products.

After the hexachlorophene catastrophe, hexachlorophene products were replaced with another microbe-killing compound, known as triclosan, marketed with all the same promises—in addition to the promise of being hexachlorophene-free—and business continued as usual. It became a common ingredient in liquid soaps labeled "antibacterial," as well as in many other consumer products—including clothing, kitchenware, furniture, and toys. For decades we poured it on our hands and down our drains, allowing it to accumulate in our water and soil.

Animal studies have shown that triclosan alters the way some hormones work, raising concerns about the effects of its use in humans. Using some antibacterial soaps might even promote the growth of liver tumors, according to a 2014 study published in the prominent journal *Proceedings of the National Academy of*

Sciences, and the culprit seemed to be triclosan. At the time, the compound was already known to have links to allergies in children and to hormone-signaling disruption that appeared to play a role in breast cancer, thyroid functioning, and weight gain.

By 2014, it was much too late to prevent anyone's exposure to triclosan, even if you knew to look for and avoid it—which most people didn't. In fact, most thought they were doing something *good* for themselves by paying extra for the antibacterial soap.

"Our interest in this was that triclosan is just so abundant," lead researcher Robert Tukey, a professor at University of California, San Diego, told me at the time. "It's really everywhere in the environment."

Because triclosan-infused products have been so widely used for many years, it is among the most common chemicals to be detected in streams. In a national health survey published in 2009, researchers from the Centers for Disease Control and Prevention found that nearly three quarters of people they tested had triclosan in their urine. Another study in 2014 found triclosan in the urine of 100 percent of pregnant women tested in Brooklyn.

"We aren't saying that triclosan causes cancer," said Tukey, drawing a very careful distinction. "We're just saying that with constant exposure, this environmental agent, which is extremely ubiquitous, can promote development of tumors."

It was not until 2013 that the FDA told producers of antibacterial soaps they needed to substantiate claims that antimicrobial cleaners have any benefit at all. The agency said in a statement, "Although consumers generally view these products as effective tools to help prevent the spread of germs, there is currently no evidence that they are any more effective at preventing illness than washing with plain soap and water."

Almost no evidence arrived from the soap producers. After a long, deliberative process, the FDA finally ruled that triclosan, hexachlorophene, and seventeen other "antimicrobial" ingredients cannot be added to soaps for consumer use because of insufficient evidence of safety (in light of copious evidence of harm), and in 2017 these ingredients were removed from the market.

The problem isn't limited to products that promise to kill bacteria. We have also exposed ourselves and our environments to preservatives with antimicrobial properties. For example, parabens are synthetic preservatives, used since the 1950s in a wide range of hygiene and beauty products—such as deodorant, makeup, toothpaste, and shampoo—as well as a lot of packaged foods. They appear on labels as *methyl*paraben, *ethyl*paraben, *propyl*paraben, or *butyl*paraben. The goal of putting them into everything was to make food and hygiene products more shelf-stable and thus more affordable and accessible to everyone in the world.

Laudable as that may have been in concept, in practice we now all have parabens in our blood. Individual products typically contain tiny amounts of parabens within "safe limits" set by the FDA, and pose no discernible threat. The concern arises from cumulative exposure over years and decades, from myriad products. Many environmental health experts have expressed concern that this could be overloading our bodies and contributing to a wide range of health problems. Though the degree of harm is impossible to know for sure, studies have found links to an increased risk of breast cancer and reproductive toxicity by way of endocrine disruption, since parabens mimic the effects of estrogen.

Parabens are antimicrobial compounds, by design. They kill a wide range of bacteria and fungi. So the question is not whether these products and practices have affected our microbiomes and immune systems, but how much those effects matter. Researchers at the National Institute of Allergy and Infectious Diseases have found that products containing parabens can block the growth of *Roseomonas mucosa* from healthy skin. This bacteria seems to help improve the skin's barrier function, and can directly kill the *Staph. aureus* that proliferates during eczema flares. In 2018, the researchers raised the concern that through this chain of events, parabens could leave people more susceptible to eczema flares.

To know exactly how parabens have changed our microbial populations, though, is impossible—both since our microbiomes are so diverse and complex, and since no one could reasonably be said to be free of parabens. Public-health advocates are pressuring the FDA to ban parabens in products sold in the U.S. The European Union did this in 2012—but the economic influence of industry on regulation in American politics makes this unlikely.

Antimicrobial preservatives like parabens have also prevented innumerable cases of food poisoning and much waste, so none of this is to say that the compounds' net effect has been for the worse. But it does serve as a warning that there can be longer-term, cumulative effects whenever we apply antimicrobial products to our skin, much less lace them into every crevice of our environments.

• • •

Graham Rook, now an emeritus professor of medical microbiology at University College London, has been leading a charge to get people exposed to—and appreciative of—biological diversity on and in our bodies.

In 2016, he, along with five other prominent immunologists and infectious-disease specialists, declared it was time to abandon the term "hygiene hypothesis." They proposed as a replacement either the "old friends" or the "biodiversity" hypothesis. The point was to emphasize that many microbes are not our enemies so much as they are just *there*, on us, probably because they play some role in supporting other microbes. They evolved along with us, and those that aren't our friends might be friends of friends, or friends of friends of friends.

The biodiversity hypothesis doesn't propose that *hygiene* is bad but that the loss of different kinds of microbes is bad—that modern inflammatory and autoimmune diseases are linked to us being deprived of exposure to the microbes we evolved to be exposed to, including pathogens as well as beneficial and neutral microbes. And we aren't just deprived of them by washing and using antibacterial products, but in all the ways we are today isolated and sterile and live in a world that is too, if you will, *clean*.

As Rook explains to me over coffee over Skype, early and regular exposure to microorganisms trains the immune system to react appropriately to threats: "It's not that children in developed countries aren't subject to enough infections when they are young, but that their exposure to the microbial world is far more circumscribed than it once was."

Maternal microbes colonize infant intestines in childbirth and immune cells are transferred while breast-feeding. Young children continue amassing microbiota in every contact with family members, while playing outside in dirt, getting licked by dogs, and sharing toys with friends. All of those microbes shape the developing immune system, which is malleable in the first few years of life, like a hot bar of freshly pressed soap.

Cesarean sections have been linked to increased risk of allergies and asthma; owning a pet can be protective against them; and antibiotic use (which kills off much more than just the disease-causing microbes) in youth has been linked to asthma, cow's-milk allergy, inflammatory bowel disease (IBD), and eczema. Cleanliness is part of the problem, Rook argues, but it's also the low-fiber diets that have changed the populations of microbes in our intestines, and the antibiotics that have changed the microbes in and on us.

Seeking out old friends does not mean exposing ourselves to dangerous infectious diseases. Small communities have sprung up in recent years advocating intentional infection with parasites and the like, to stimulate the immune system and ideally to treat autoimmune conditions. It's an interesting idea, but not endorsed by any official medical entity: the risks are definite, and the benefits hypothetical. Likewise, I don't think anyone should sneeze in anyone else's face. But people know this, and inhaling a sneeze isn't the main way we get respiratory viruses. The flu and other viruses kill millions of people globally, and much of that could be prevented by cutting off chains of transmission with simple hand washing.

But there are harms associated with too much isolation and removal, and with overusing soaps and antibiotics. The best advice right now is to think of hygiene as similar to medicine—extremely important in some scenarios, and also very possible to overdo. The same goes for exposure to microbes. Historically, exposure has been a much bigger danger than over-cleaning. Now, in much of the world, it's the reverse. So what's a healthy amount of exposure, and how do you achieve it without compromising safety?

Jenni Lehtimäki is a rising star in skin microbiome research. Her research focuses on elucidating the ways microbes mediate our relationship to our environment.

Her group of researchers at the University of Helsinki in Finland was among the first to show evidence that microbe populations on the skin can have more-widespread effects on the body. When Lehtimäki exposed mice to the bacteria *Actinobacteria*, she was able to track a small but measurable influence on immune responses.

Lehtimäki is originally an ecologist and evolutionary biologist. She started focusing on skin microbes when it became clear to her that they relate to allergies, and to the environment where you are living. In a recent study, she and colleagues looked at dogs living in urban and rural environments in Finland. They found that the dogs that lived in rural environments had a lower risk of allergic symptoms. But it wasn't just location that seemed to matter. Lifestyle factors—like how much time the dogs spent outside, and whether they spent time with other animals—were also associated with allergy protection.

To Lehtimäki, the biodiversity hypothesis means microbes are the fundamental actors in how our immune systems are shaped. Early studies are showing that it can be very difficult to permanently change your microbiome, but that everyone is affected by temporary exposures. They stimulate our immune systems through contact with our skin and the lining of our guts, even if they do not permanently stick with us.

In practice, these findings suggest that maintaining healthy exposures—and a healthy immune system—is an ongoing and active process. It is very unlikely that anyone will be able to live an isolated life, barely leaving their purified-air skyscraper, and maintain a diverse microbiome by taking a daily pill. Lehtimäki's research helps explain some lifestyle factors that we've long known to be associated with good health, but weren't sure why. Being exposed to nature—having pets and living communally and engaging with the natural world—affect our own microbiomes, essentially becoming an extension of it.

People you live with also share your microbiome. Cohabiting partners start to have similar biomes, according to a 2017 study from the University of Waterloo that swabbed the skin of couples who lived together and separately. Using machine learning, lead researcher Josh Neufeld was able to pick out cohabiting couples based on their microbiome profiles with 86 percent accuracy. He told me that foot microbes tended to be the most similar—probably because we step on the same microbes that settle on the floor. People who cohabitate generally have more microbial diversity—as do people who have pets, drink less alcohol, and exercise more.

As Mark Holbreich, the coauthor of the study among the Amish, pointed out to me, "If you look at aboriginal tribes, their microbiomes are very different from urban dwellers. And if you move an aboriginal person into a city, within days their microbiome changes. The whole family changes, the house changes—the microbiome is a very vibrant sort-of organ."

Lehtimäki is working on modifying the microbiota in our homes and offices in order to *increase* microbial exposure. Her approach is pragmatic: "Because people are lazy and they don't want to do much, maybe you just bring the microbes to the home somehow." Some researchers have tried to transport "good microbes" into people's houses and apartments using carpets.

There are also those carpets with the capacity to walk and love, known as dogs. Lehtimäki's study of dogs in Finland found that when a dog was allergic to something, its owner was more likely than the general population to be allergic as well, suggesting a shared microbial environment.

Increased exposure to natural environments also seems to have a broader effect on health. A number of studies have reported associations between green-space exposure and self-reported health, birth outcomes, and reduced morbidity. A 2018 meta-analysis found statistically significant associations between exposure to green spaces and reduced blood pressure, heart rate, cortisol levels, incidence of type 2 diabetes, and death from cardiovascular disease.

Exercising outdoors may also have health benefits you don't get at the gym. Much work has been done in this area by Diana Bowler and colleagues in the UK, who compared the effects of exercise in "natural" and "synthetic" environments and found that a walk or run outside "may convey greater health benefits than the same activity in a synthetic environment." In another meta-analysis of walking groups, outdoor walkers were found to have significantly improved blood pressures and body-fat percentages relative to indoor walkers, leading researchers to conclude that "walking in a green space or natural area may offer health benefits above walking in an urban environment or on a treadmill."

Various theories have been offered to explain these findings, including the emotional boost you might get from spending time outside. But Lehtimäki is particularly interested in the role that exposure to microbes may be playing in mediating this effect. The scientific understanding is still very limited. In the meantime, she tells me that she does what she can to expose herself to nature—and limits her use of antibacterial products. She rarely uses deodorant and avoids hand sanitizers. She showers in the same way as most of the other microbiologists, which is to say, conservatively.

VII

VOLATILE

The hatch popped open, but Claire Guest's aging golden retriever Daisy didn't do her usual instantaneous leap out of the car. Instead she just sat there, her head cocked sideways, staring up at the young scientist.

"She was a bit wary of me," Guest recalls, "like something was bothering her."

It was a sunny afternoon in 2009. The day had started out with a routine trip to the park near their London home. "She looked up into my eyes, and I said, 'What's wrong?'"

Guest, thirty-three at the time, was working as a medical researcher in an obscure field: studying the long-held but poorly understood idea that dogs have the ability to smell cancer. She had read anecdotal reports, throughout history, of pets acting differently around owners who become sick. As a biologist she wanted to know more about what it was the dogs could actually be detecting.

Daisy was part of the research program. She had been living with Claire for the past year as part of a nonkennel policy, in which all the dogs go home to various volunteers and families.

Seeing Daisy's reaction, Guest was jolted out of the moment. A few days ago she had felt a little lump in her breast, but she was young and ignored it. All of a sudden it clicked. The biopsies would come back as breast cancer.

Of course, Claire had never expected Daisy would find a tumor in her. When she did, it turned her novel academic interest into an all-consuming mission to understand what may be a major missing piece in modern medicine.

Guest, whose cancer is now in remission, has gone on to work full time with dogs who can pick up on signs of cancer and other conditions. She founded a research organization called Medical Detection Dogs. The organization has a bio-detection department looking at detecting particular diseases from a swab or sample for an individual, and an assistance department training dogs to live and work with individuals and to give a warning in a medical emergency.

"The struggle was to get this treated as a true science," Guest says—to get people to accept "a biosensor with a fluffy coat and a waggy tail." Compared to humans, dogs have far more sensory receptors and a bigger part of the brain dedicated to olfaction. If we had similarly structured brains, we would be more aware of the thousands of volatile chemicals that constantly radiate from each of us. These are what the dogs detect.

Called "volatiles" for short, the full scientific term is volatile organic compounds (VOCs), meaning simply carbon-containing chemicals more complex than CO_2 that are suspended in air. They come along with everything that comes out of us, from breath to mucus to urine, and are even produced by the normal functioning of our skin. These together form a sort of chemical fingerprint that's unique to each of us—our "volatolome" (like genome or microbiome, the "ome" being lately used to denote any large set of things).

Subtle differences in the volatolome are what allow dogs to pick a single familiar person out of a crowd. It doesn't take a major change in that specific volatolome for a dog to detect that something is awry. And it is becoming clear that our bodies produce chemical signals that reflect states of health and disease. Particular diseases can have their own unique patterns of influence on the volatiles we emit, and dogs can be trained to find these like any other odor, even when they're totally unfamiliar with the person's usual odor signature.

For example, dogs have proven helpful in detecting high blood-sugar levels in people with diabetes. They have also had success detecting Addison's disease, the autoimmune condition in which the immune system attacks the adrenal glands. This causes a person's cortisol levels to plummet, throwing off vital metabolic processes. The inverse scenario, elevated cortisol levels, is also seen in many diseases. By detecting that, Guest hopes, the dogs may also be able to pick up high-stress states that could precede a panic attack, or even a heart attack or stroke.

"In the UK the medics tend to be very skeptical of it all," she says, "but when they come in and watch the dogs and see it happen, it's quite powerful."

Detecting the smell of Parkinson's disease has also shown promise—and not just in dogs. Guest has spent so much time training dogs to detect the disease that she believes she can smell Parkinson's disease herself. Indeed, some nurses swear they can smell cancer at late stages. Part of the reason this is difficult to believe is that, until very recently, scientists had no idea exactly *what* people would actually be detecting. This is where the new science of the skin microbiome could play a role.

A century ago it was reported that Parkinson's disease was associated with changes in the sebum of the skin. Such changes could lead to shifts in the microbiome and, hypothetically, the volatiles produced by the ecosystem. I ask Guest about this theory and she lights up: "I'm fascinated by the skin microbiome!" She says she believes it will be central to future research.

Just how microbes of the skin and mouth are contributing to the volatolome is difficult to know, but their production is almost certainly due to a mix of by-products of human metabolism that we excrete, and which are then further metabolized by microbes. The chemicals we emit are a product of both microbes and our own bodily functioning.

"It wouldn't surprise me if it isn't so much that Parkinson's has an odor as that the change in neurotransmitters leads to a change in the microbiome, and that leads to a different smell," says Guest. "It wouldn't surprise me at all if what the dogs are picking up is actually a change in bacteria that's associated with disease."

As evidence supporting the idea has begun to grow, Guest has started to be approached by researchers who think it's at least worth a shot. This is especially true for diseases that could be treated if only they were caught earlier. In 2018, for example, researchers in Mexico analyzed used menstrual pads and found that the volatolome of the female genitourinary tract changes in predictable ways when a person develops cervical cancer. The chemical output is likely the result of changes in the vaginal microbiome that may be secondary to—or possibly even a cause of—the disease.

Such changes in chemical output may not be detectable by human noses, but machines can detect them. This is almost certainly what dogs respond to in such cases.

"This concept has long been dismissed as an old folklore tale, but it's not," says Steve Lindsay, a public-health entomologist at Durham University in the UK. "Dogs are sensing much more than our body language or disposition; the dogs really can discern health from disease based on chemicals we emit. Sometimes the dogs do this better than the best tests science has."

Lindsay studies how insects affect our health. He became curious about the odors that human skin bacteria emit while studying the chemical signals that mosquitoes use to communicate and to locate and infect humans. Among the most worrying challenges for him and colleagues is that after more than a decade of dramatic reductions in malaria infections and deaths, over the last two years the world has experienced a slight *increase* in both.

Malaria spreads through a complex cycle in which mosquitoes transmit parasites to humans and then humans transmit them back to mosquitoes. Scientists developing tests to detect malaria—which is caused by the transmitted parasites—are suffering setbacks because the parasites are mutating. Some of the new types of malaria no longer produce the specific protein that tests are designed to detect. Stemming an outbreak hinges on detecting asymptomatic carriers who seem totally healthy but can still pass parasites to the local mosquito population.

At the 2018 meeting of the American Society of Tropical Medicine and Hygiene in New Orleans, an international group of researchers presented findings that would once have been thought absurd. A team of British researchers working in Gambia checked hundreds of schoolchildren for malaria parasites and gave them all a pair of socks to wear overnight. The team then collected the socks and sorted them based on which children were infected, then shipped them back to London and kept them in a freezer for several months.

Researchers then mixed the socks of the kids who had malaria with those who didn't. They showed the socks to dogs from Medical Detection Dogs, who then sniffed each sample. If a dog thought it detected malaria, it would freeze over the sock. If not, it would move on. The dogs correctly identified 70 percent of the socks belonging to infected children, and were even able to detect infected children with lower numbers of parasites than are required to meet standards for rapid diagnostic tests set by the World Health Organization.

These dogs are not about to replace standard blood tests, but the evidence that we give off a detectable chemical signal when we're infected by malaria—and, presumably, other infectious agents—was a breakthrough. As for how that happens, Lindsay posits that the distinct compounds emitted may be due to alterations of the skin microbiome. He also notes that even infected blood in a dish in a lab gives off different chemical signals than prior to being infected with the parasite.

Eventually, he thinks dogs could be a serious way of diagnosing people who don't show any malarial symptoms, but are still infectious—though, Lindsay says, researchers are limited by some people's aversion to dogs. There are also critical cultural considerations, like that dog saliva is considered "dirty" in many Muslim cultures, possibly due historically to transmission of rabies. Indeed, dog saliva is, like human and other saliva, loaded with microbes that help dogs digest food and maintain healthy oral microbiomes that protect their teeth, and that can also easily get into the bloodstream of anyone they choose to bite. This makes biting an especially dangerous attack; even if the only initial injury is some minor puncture wounds, a person can die within a few days without antibiotics. Animals bites are taken very seriously in emergency rooms around the world. As Lindsay put it, you don't want to be the white fellow who just waltzes into an African village with a dog and expects people to be grateful.

A more promising notion is that these dogs could be deployed at entry points to countries where malaria is nearing eradication. They could hang out at airports and ship terminals and train stations and detect people carrying malaria parasites into places like the Zanzibar Archipelago, where efforts to eliminate malaria are complicated by a steady stream of visitors arriving from the mainland.

The longer-term goal for scent-based diagnostics would be to let dogs live their lives and to find ways to use "electronic noses"—or "eNoses," as they're called by people who find they are wasting too much time saying "electronic noses." In fact the existing prototypes look nothing like noses; they look more like credit cards, which makes the name especially strange because it's not intuitive that a "nose" would be used to detect cancer or malaria. But my mind is in product-marketing mode and I'm getting ahead of myself. Before these products could ever be marketed or sold to doctors (or directly to patients), scientists must figure out exactly what it is the dogs are detecting.

It's not just our skin chemistry that changes, but our breath, which gets most of its smell from the microbes in our mouths and throats. When the malaria parasite invades, it somehow alters compounds humans naturally exhale (or otherwise give off). At the 2017 tropical medicine and hygiene meeting, bioengineers from Washington University in St. Louis reported the discovery that people with malaria exhale a distinctive "breath print," and used that to develop a breath test that caught 83 percent of cases of malaria in a preliminary test, which they used to diagnose children in Malawi.

The bioengineers reported that they had found malarial infection was associated with abnormal levels of six different compounds that are normally detectable in human breath. This seems to indicate that the parasite isn't just changing one signature metabolic process, but throwing a whole system out of balance.

They also found something unexpected in the breath of children with the parasite: two types of compounds known as terpenes, which are typically linked to strong odors exuded by plants like pine trees and other conifers. One of the terpenes is known to be produced by plants that attract mosquitoes to feed on their nectar. The researchers believe that, in a brilliant strategic move, the parasite may be "hijacking" the mosquitoes' preexisting attraction to the odor to encourage them to bite infected humans, thus extracting parasites and facilitating the spread of disease.

"The terpene is probably a survival mechanism for the parasite," Audrey Odom John, a professor at Washington University School of Medicine, said at the time. She also suggested that the compound "might be useful in boosting the effectiveness of mosquito traps."

If compounds we emit—based on the microbes we carry—can attract mosquitoes, the implications go beyond malaria. Our chemical signals may even help answer the age-old question of why some people can sit around the campfire and get eaten alive, while others are barely bothered. The current approach of slathering ourselves and our lawns with toxic chemicals is begging to be improved. Some researchers believe the answer lies in the microbes on our skin and in our mouths—not killing them, but masking the specific compounds they give off that mosquitoes detect.

Researchers at Texas A&M University, for example, found that modifying the *Staphylococcus epidermidis* on people's skin can put us into a sort of stealth mode in which mosquitoes can't find us. This was done through a complex process of modifying the chemical signals that the bacteria give off. At least as proof of concept, this mode of thinking could transform the insect repellent industry. As entomologist Jeffery Tomberlin put it, "We might want to modify the messages that are being released that would tell a mosquito that we are not a good host, instead of developing chemicals that can be harmful to our bacteria on our skin, or to our skin itself."

• • •

If so much of what we've learned about human evolution says clearly that we are a social species—that we rely on others for survival, and that our individual faults can often be explained as useful in the context of a community, where a diversity of skills and assets is actually better than having everyone with perfect SATs and no one who can fix a toilet—why would we evolve to smell *bad*? To actively repulse other people, to drive them from the room? Even when we aren't sick at all?

The argument against the ability to achieve a smell-free steady state is that the odor-producing bacteria are there because they play some useful role in our existence. We didn't evolve *to smell*, we evolved in harmony with microbes that serve a function for us—and also, unfortunately, happen to sometimes produce bad odors.

Consider how Rob Dunn, the professor of applied ecology at North Carolina State University and a coauthor of the study about skin mites, explains feet. As a human with a functioning nose, he, of course, agrees that the smell of feet can be one of the most repellent things about a body. The stench would be evolutionarily indefensible, unless it came with some arcane survival benefit, like at some point we were using our smelly feet as weapons against enemies. I've found no historical record of this. So Dunn urged me to critically engage with the question of why feet smell.

In other animals, foot odor seems to directly serve a purpose. Bumblebees, for example, emit odors from their feet that are unique to each individual bee. These odors mark their tracks so that their cohorts can follow the stinky footprints, leading them to one another, or to food.

If human foot odor serves no such appetizing or prosocial purpose, the bacteria that produce the odor may be common because they serve some other useful role. One possibility, Dunn notes, relates to the fact that humans walked barefoot until very recently, and so were susceptible to cuts and scrapes on the feet that could become infected. Before antibiotics, minor infections like that were often fatal. While fungal infections like athlete's foot are typically low-threat annoyances, a break in the skin can let the fungi into the bloodstream and cause havoc. So it could be evolutionarily adaptive to harbor innocuous species on the feet that might help prevent an infection.

Some bacteria are even known to produce compounds that have antifungal properties. One species that is commonly found on our feet, *Bacillus subtilis*, generates compounds that are lethal to fungi of the sort that tend to cause foot infections like athlete's foot or toenail fungus.

Unfortunately, *B. subtilis* also smells terrible. Dunn has traced much of the characteristic "hot trash" smell of stinky feet to a compound called isoflavic acid, which is produced when *B. subtilis* metabolizes the amino acid leucine in our sweat. Compared to the rest of the body, the sweat that comes out of our feet has especially high levels of leucine. He posits that this could be the result of coevolution between ourselves and our skin bacteria.

This specific example is still hypothetical, but the basic idea is that compounds like leucine are not shooting out of our feet without playing some role there, and nor are the *B. subtilis* there purely to annoy and embarrass us.

Our feet may have evolved to produce lots of sweat, containing leucine, to feed specific bacteria that kill fungi, reducing our risk of foot infection. So while rancid feet may create an impediment to finding a sex partner, people thus afflicted were also reproductively *advantaged* over those who died of septic shock from a foot fungus.

This model of our skin, its secretions, and its microbes as a symbiotic ecosystem raises the question of how much we should be washing off. In Dunn's theory, anything we do to make bacteria with useful metabolic abilities similar to *B. subtilis* less abundant (as in cleaning) could increase our risk of fungal infections. Abnormal foot microbiomes may even help explain why such fungal infections are so common today. At the same time, no one wants to smell terrible. So the question is just how to strike the right balance.

As with any function of our bodies, the effect of smell is not some binary situation, where we either have "body odor" or don't. More likely we have milieus of smells that present on various scales in various situations, and through which we may express ourselves in no less complex a way than we do with the intonation of our voice or the subtle contortions of our face. Many people tell me they think their romantic partner smells good—referring to that person's steady-state odors that social standards don't permit most of us to experience with most other people.

In learning all this I was curious what these airborne compounds (and their smells) are all about—what we might be gaining by having a human smell, and what we might be losing by washing it off. What if all the soaps and colognes and perfumes we use—no matter how "natural" they claim to be—are also changing and masking signals that serve some purpose? The hundreds of subtle volatile chemical signals we emit may play roles in communicating with other people (and other species) in ways we're just beginning to understand.

• • •

Chemistry between people is not just romantic, and not just about signaling health and disease. There is something to physical presence that cannot be replicated on screens and in texts.

According to innumerable magazine covers and books and academic papers, the sense of being isolated and disconnected is a defining one of our era. In the same way that coffee shops draw us even when we just sit in front of our laptops and endure bad music and don't acknowledge anyone else—except when asking strangers to watch our laptops while we go to the bathroom—something about even the small contacts and physical presence of others seems to sustain us. This might be, in part, due to the chemicals we all give off.

Ben de Lacy Costello has studied the volatiles found in human feces, urine, and saliva, so that you don't have to. He explains that stress and anxiety have been shown to have clear effects on the chemicals we give off. (This could be an important confounding factor if making a disease detection device. *Disclaimer. Do not use while anxious. This may produce a false positive test result, which will only make you more anxious, and so could precipitate a vicious cycle of stress that could actually kill you. Oh, great. You're worrying about the stress now, aren't you? Just forget I said anything.*)

I came across Costello's work in 2016, when I interviewed him for a story I wrote about contagious emotions. The story was inspired by a recent study by climate scientists who had set out to understand whether human breath—which is, after all, enriched in carbon dioxide—contributed to climate change. The lead researcher, Jonathan Williams, is an atmospheric chemist with Germany's Max Planck Institute for Chemistry. When Williams studies the climate effects of gaseous emissions from plants and animals, he uses finely calibrated machines that sense the slightest changes. So his team brought these sensors into one of the most volatile environments in the world: a European soccer stadium.

The amount of carbon dioxide the scientists detected was surprisingly inconsequential, but in true scientific fashion, something much more interesting showed up on the sensors. When Williams told me this in an interview for the story, I immediately asked him if it was alien life. He said no, but other strange chemical signals seemed to be coming from humans. They would come and go, at various points in the match. As Williams sat and watched the fluctuating readings on the air sensors, he got the idea that they might be related to emotions.

In the course of a soccer game, the crowd goes through stages of elation and anger, joy and sorrow. So Williams began to wonder, as he put it to me: Do people "emit gases as a function of their emotions"? Possibly to communicate with one another? And with other species? If we do, it wouldn't be unprecedented. Plants are constantly emitting volatiles, of course, from the scent given off by a bouquet of roses to far more subtle signals. Plants are well known to release chemicals after they have been "attacked" by an animal trying to eat them. Long known as "herbivore-induced plant volatiles," these served, scientists thought, to warn adjacent plants about predators in the area.

More recently, researchers have learned that the signals plants give off to one another are myriad, communicating about both threats and resources, and are overlaid into an ambient "infochemical web." The functions go well beyond textbook examples like flowers attracting bees. Even trees are giving off compounds to convey information about their identity. Tear some leaves off a tree, for example, and it will emit chemical signals.

The grounding effect of walking into a forest can be in part due to the change in the air that's hitting our airways and skin. The air we describe as "fresh" might be more than just clear of contaminants—the air pollutants responsible for seven million premature deaths every year—but also laden with chemical signals from plants and animals. *Fresh* air means more than just the absence of bad things; it means the presence of good. This could partly explain the health effects that researchers have associated with being outside.

The idea of airborne "pheromones"—chemicals that specifically influence mating behaviors—tends to be dismissed as pseudoscience. The concept has certainly been distorted in attempts to peddle human attraction in an aerosol can. One online review site for pheromone products describes a spray called Pherazone for

Men as "best for attracting women," while another called Nexus Pheromones is "best for getting laid." Then, of course, there is TRUE Alpha, which is "best for trust and respect." (Because when you want someone to trust and respect you, the best approach is to trick their brain with chemicals.)

While I have not personally field-tested these products, there's no single compound that will make any person's eyes turn into hearts à la Bugs Bunny. But the basic concept of chemical attraction is supported by the existence of the volatolome. Dogs and most of the rest of the animal kingdom can detect some chemical signals from an ovulating female hundreds of yards away, and humans are not likely exempt from the practice of emitting chemistry that corresponds with hormonal changes. Though most volatile chemical signals seem to be imperceptible to our relatively humble noses, our overall mélange of gaseous emissions clearly signals more to other humans—in contexts sexual and otherwise—beyond whether or not we smell attractive or repulsive.

Costello believes the number of chemicals in the volatolome is likely in the tens of thousands. Signals between individuals could involve trillions of permutations, accounting for the subtle individuality of the smells in our armpits and on our breath, and every other part of the body. Whether or not these airborne concoctions can ever be reproduced and bottled to induce love, what's clear is that the compounds we emit are not incidental. This is at least reason to question the wisdom of repressing them.

. . .

While writing this book, I spent two weeks working in an addiction clinic in Connecticut, seeing patients under the guidance of actual addiction specialists. The relatively new field of addiction medicine is now largely focused on opioid addiction—endeavoring to treat the catastrophic effects of what often started out as medical treatments. I mostly saw patients who were on anywhere from their first to their hundredth shot at getting—the word they all use—"clean."

The word in this context feels more apt than anywhere else I hear it. In the addiction clinic, *clean* is an encapsulation of all its meanings, from removing toxic contaminants to seeking spiritual purity. The goal of treatment in addiction is not simply to quit using a substance, but to actively build a new life without it. This requires intensive, constant work and vigilant focus. Many people find it helps to see the process as a renaissance—an opportunity to be reborn, to reconceive of oneself entirely, and to start again.

At the state-funded facility where I worked, in a state hit especially hard by the opioid epidemic, there were almost no windows, and the main activity was sitting quietly in a room with a small TV that not many people were watching. Most of the residents were there by court order, arrested on minimal charges related to procuring narcotics or procuring the funds to procure narcotics.

The time in rehab can be intensely boring. But the real challenge was remaining clean after leaving the house. If you go back into the same social environment, with the same people who enabled your addiction before, the odds of relapse are almost 100 percent. If you do not have a concrete plan of exactly where to go and what to do instead of using, relapse is all but guaranteed.

Being *clean* in this sense requires the opposite of isolation or barrier-building; it requires opening oneself to new exposures. This mostly means new people: building deep, meaningful, honest relationships. This is where the state-funded program had to turn people out on their own. Programs like Narcotics Anonymous were available to provide ongoing community and mentorship, and they tend to have good results for some people, but they require the sort of radical honesty and commitment that addiction has spent years rewiring the brain to avoid.

For the rest of their lives, even decades after quitting, even if they never touch so much as a single cigarette, many people will still understand themselves as *addicts*, and this understanding will guide their continued abstinence. Getting clean only works, many addicts told me, when this new identity is part of a new approach to life; one repopulated with new people and hobbies and habits.

Behavioral science is clear: to *stop* any old habit is difficult and often a failed endeavor. The motivation is effectively drawn from the motivation to *start* doing or being something else. As with the antimicrobial approach to skin hygiene, it does not work to simply remove things. Thinking about *getting clean* as a monastic, solitary, painful errand of elimination and deprivation is unsustainable. Seeing it as a process of embracing change and creating relationships is a far more effective path.

Justin McMillen sees promise in this concept for addressing many modern health epidemics. McMillen is a square-faced and square-shouldered athlete with a crew cut and short beard who grew up "in a lumberjack environment" and who can dive sixty feet on a single breath. As a young carpenter in Los Angeles, he began using heroin. During the collapse of the housing market in 2008 he lost most of what he had, and he was living in a garage when he hit rock bottom.

Through years of athletic competition, McMillen had discovered that his body could be pushed to extremes. The challenge and pain made everyday life—working as a carpenter and living in a comfortable apartment—seem boring. This was the stability he had long thought he wanted. But when he wasn't pushing his mind and body, he felt he needed to stimulate them in other ways, by ingesting or injecting whatever might stimulate that dopamine circuit that was begging to be used again.

He came to conceptualize addiction as something that rewires the prefrontal cortex, shifting reward structures, partly based on studying the work of neurologist Dan Siegel, cofounder of UCLA's Mindful Awareness Research Center. Siegel's work emphasizes the importance of the prefrontal cortex in interpersonal connection. "As the PFC becomes dysregulated, that makes it harder to connect," McMillen explains. The isolation makes the mind more desperate for stimulation. "It's a vicious cycle."

As McMillen started to recover from his own addiction, he noticed that isolation seemed to be particularly striking among men. In Portland, Oregon, he launched a small addiction program for men called Tree House Recovery, based on teaching them to connect. It is predicated heavily on physical contact. A "physical empowerment director" works in tandem with a more traditionally focused clinical director overseeing exercises meant to create trust and connection between participants—to create scenarios where people must depend and rely on one another.

Though residents live in a house that looks just like any other in Portland's sea of craftsman bungalows, the program is considered a partial hospitalization facility and is covered by many forms of health insurance. In promoting it McMillen has become one of the rare public advocates of physical touch among men, touting health benefits like "lowering blood pressure, strengthening the immune system, improving memory, reducing pain, and more." When he demonstrates this practice in local news segments with male anchors and reporters, the level of awkwardness varies from moderate to high. McMillen emphasizes that touch can be as simple as a pat on the back; he's not expecting that men will start holding hands with acquaintances.

"Coming out the gate and saying, 'Hey, everyone's gotta hug each other,' that likely wouldn't have worked," he tells me. "Touchy-feely" is often seen as a pejorative in the recovery genre. So McMillen has found, through years of trial and error, that the key to real platonic touch is to get people to choose to do it themselves without much thought—which tends to happen as a by-product of first doing it in some familiar framework.

Since these norms seem to melt away in athletic competition, especially wrestling or boxing, sports are a way to teach men that touching other men is good and okay. But of course, McMillen wasn't going to have the men in his program actually hit one another. So he developed what he calls "action-based induction therapy," which looks like mixed martial arts, but the point is just to get the men to experience platonic touch. "It's in no way a situation where they're beating each other up," he reassured me, as my mind drifted to a *Fight Club*–type scenario for men seeking to learn to feel again.

"We can develop trust through mirrored movements, promote physical touch in a way that's comfortable because it's 'masculine' enough," he says. "After class you see guys hanging off each other, the social ideas of boundaries kind of drop away."

Part of the reason we greet with handshakes and hugs is the universal knowledge that breaking physical barriers instantly makes other barriers more permeable. I experienced this while reporting on a recent "wellness festival" in Palm Springs, where a notably high proportion of attendees identified as addicts. In one session we were asked to stand in two lines facing each other, with our faces only about ten inches apart. We were told to never break eye contact, and then to talk about our most serious sources of anxiety. And initially it was as awkward as it sounds, but something about the physical proximity and alignment was like taking a kink out of a hose. In one minute I gushed as much to a stranger as I would have in an hour of talking to a friend—and with a friend I'd probably be letting my eyes wander, and crossing my arms, and doing all sorts of other subconscious things that psychologists would tell me are actually about blocking connection.

The health benefits of touch itself—platonic touch devoid of any sort of relationship—is well documented. In 2019, I interviewed a pioneer in the research, Tiffany Field, a developmental psychologist who went on to found the Touch Research Institute at the University of Miami's Miller School of Medicine. Field has spent decades trying to get people to touch one another more. Her efforts started with premature babies, when she found that basic human touch led them to quickly gain weight. They averaged fewer days in the hospital and $3,000 less in medical bills.

This led to documenting effects of "touch deprivation" on kids: it has been found to lead to permanent physical and cognitive impairment, and to social withdrawal later in life. Field has published similar findings about the benefits of touch in pregnant women, adults with chronic pain, and people in retirement homes. Physical touch isn't known to make adults grow larger, but as little as fifteen daily minutes seems to have myriad benefits.

In a more recent study that made headlines about hugs helping the immune system, researchers led by the psychologist Sheldon Cohen at Carnegie Mellon University isolated 400 people in a hotel and exposed them to a cold virus. People who had supportive social interactions had fewer and less severe symptoms. Physical touch (specifically hugging) seemed to account for about a third of that effect, the researchers concluded. The mechanism is unknown, and guesses often center vaguely around touch receptors leading the brain to release endorphins and other chemicals that bolster the immune system. An equally compelling hypothesis might be that people who touch are sharing microbes, and that this is at least partly responsible for any effects.

I may be eager to buy into this idea because it partly explains something that I've experienced. Like a lot of jobs, being a writer today tends to mean a lot of digital communication—spending whole days emailing and interacting on Twitter and texting and talking to people on screens. The commonly voiced modern affliction is that we're processing images and language in ways that simulate connection almost constantly, and yet we can feel more alone than if we spent the day with just one actual person. The touch and exchange of chemistry surely can't explain that effect on their own.

But whether the reported benefits of physical contact come from the sensation of touch, or the chemical signals animals send into the air, or the microbes we share whenever we're near others, we would do well to see our bodies as part of a community—stronger together than they can ever be alone.

VIII

PROBIOTIC

Not far past some blocks of boarded-up row houses in Baltimore, the horizon suddenly fills with food carts selling collagen drinks and flocks of athleisure-clad influencers streaming into the city's palatial convention center. For four days it will be transformed into the world's premier wellness trade conference: the annual Natural Products Expo. If storefronts and wellness retreats and festivals are for consumers, this is where retailers and distributors go to stock their catalogues for the next season of wellness trends.

As at the Indie Beauty Expo, the sellers at the Natural Products Expo are linked by a term that has no agreed-upon meaning. Boutique brands selling Himalayan sea salt and charcoal toothpaste occupy booths next to major retailers of oat milk and collagen powder. LaCroix has an enormous booth, as does mac and cheese purveyor Annie's—an example of marketing triumph, where a company was able to take a fifty-year-old product (Kraft mac and cheese), repackage it with slightly different ingredients, and sell it to concerned parents for twice as much because it was labeled "natural." Dr. Bronner's is also represented, offering a new hand sanitizer consisting of alcohol, water, glycerin, and peppermint. The bottle reads "99.9% effective against germs." It seems my ideas about embracing skin microbes did not inspire David to break the mold. I'd thought we were connecting. Do you ever really *know* anyone?

I first attended the Natural Products Expo four years earlier, and the change since then is dramatic. The presence of probiotic products for the skin—in addition to the gut and mouth and vagina—represent a category that previously barely existed. Now a market is exploding around a concept that upends the central tenet of the hygiene revolution: the idea that adding bacteria to your body can prevent or reverse all manner of disease.

At a booth for a company called Just Thrive, a man looms over me clutching a jar of pills. Billy Anderson's polo shirt is tucked into his jeans. He is a retired pharmaceutical salesman-turned-executive with the bearing of a former collegiate baseball player, which he was. It is late in the day and he sounds like he is going through the motions as he begins his sales pitch:

"These bugs used to be found abundantly in the environment, in the dirt we lived in, the food we ate, the water we drank," he tells me. "But because we farmed the same land over and over, and because of pesticides and herbicides and antibiotics, our soil got depleted of microbes."

Here he veers into uncharted territory, implying that his pills could treat, among other things, autism. "Parents will take their children to the doctor," he says, "and the doctor will go, like, *Holy cow, this is awesome. What'd you do? The numbers are—things are looking really good*. Parents say, *I gave them Just Thrive*."

Anderson and his wife, Tina, who also formerly worked in the pharmaceutical industry, both quit their jobs and, according to their company website, "sought out to find a probiotic that was nature's true probiotic." The label is a master class in selling what something is not: "non GMO, and made WITHOUT [emphasis theirs] soy, dairy, sugar, salt, corn, tree nuts or gluten." As with so many products at the expo, it is "vegan, paleo, and keto friendly." A small bottle sells for $49.99 (plus $4.99 shipping).

What's less clear is what the product *does*. Seen on other products at the show and on store shelves, the term "probiotic" seems to be treated as a synonym of "good for you." Claims vary from treating complex neurologic symptoms to general self-maintenance. Nearby vendors are selling probiotic house cleansers and probiotic deodorants. Elsewhere in the expo space, an enormous sign hanging above the crowd reads "THE TRUE PROBIOTIC," and under this, in pink script, "for women." The product label says it is a "vaginal probiotic" that "promotes urinary tract health."

When I reach for a sample the salesperson says, "No, no," and pulls the basket away. "You want the regular ones." She is too slow, though, and I have my two-pill sample package of Jarro-Dophilus Women. I later took them without reservation (they had no noticeable effect), because they were an *oral* probiotic. Regardless of whether a person has a vagina, swallowing bacteria does not send them to the vagina. The only way that bacteria could hypothetically travel from your gastrointestinal tract to your urinary tract or vaginal canal would be if they got into your blood. This would be an urgent medical emergency.

The bacteria in this pill were all *Lactobacillus*, which is the genus predominantly found in yogurt. Though yogurt is certain to contain living bacteria, pills like these and other probiotics are more variable. Most bacteria are not shelf-stable for long, which is why probiotic products often require refrigeration (Jarro-Dophilus

Women did not). It can be difficult to know how many living bacteria are truly still alive in any probiotic supplement—much less how many are making their way through your stomach acids and sticking around in your bowels. Probiotic dietary supplement labels are required to list the amount of viable bacteria contained in the pills or capsules, but this can be challenging to state accurately since, unlike most ingredients in food and drugs, this ingredient is alive.

By definition the bacteria must be alive in order for a product to be called a probiotic, according to the FDA. Kombucha is a probiotic, for example—you can see the microbes floating around right there at the top, actively fermenting the sugars in the tea into alcohol. The glob is known as a SCOBY (symbiotic culture of bacteria and yeast). These microbes continue to actively ferment the sugars in the brew even while it sits in the refrigerator, meaning that alcohol levels in the final product can sometimes vary widely. Keeping the microbial activity consistent from bottle to bottle has been a challenge for brewers and regulators, and occasionally leads to batches with far higher than the anticipated amount of alcohol.

Other live-bacterial products pose similar challenges for producers and users. New preservation methods such as freeze-drying (lyophilization) have shown promise in helping to deliver a consistent product, though microbial preservation and delivery techniques aren't standardized. To further complicate things, some products that claim to be "probiotic" actually include "bacterial lysate." That means the bacteria have been lysed, or heated, killed, and broken down.

It's unclear what the effects of ingesting or applying lysates may be, but it's surely different from using a living organism. Researchers tell me that it's hypothetically plausible that these dead bacterial parts could have some effect on the immune system. After all, dead parts of viruses are used in vaccines to stimulate the immune system. But expecting *probiotic* effects from introducing lysed bacterial parts into your existing microbiome is about as plausible as expecting bacon to populate your pig farm.

The term "probiotic" was beginning to appear at the Indie Beauty Expo, too. The company LaFlore offered me a "probiotic cleanser" ($42) and "probiotic serum concentrate" ($140). The products consist almost entirely of oils and herbal extracts common to those in adjacent booths, but near the middle of the ingredient list there is some *Lactococcus lysate* as well as fermented *Lactobacillus*. The effect of such ingredients is unclear—in the scientific literature and in my subsequent experimentation with the product. But LaFlore makes no definitive claims about what those bacterial elements are supposed to do for my skin, and the proprietor was very kind and did wear a white lab coat. She let me mix my own serum in a glass bowl to show how simple and natural the ingredients are. Seeing colors form as I added various powders was mesmerizing. It reminded me of mixing primary-color paints in kindergarten, and watching the red and yellow become orange, which still always feels a bit like a magic trick. Someone from the company was filming this, presumably for Instagram.

Another company at the Indie Beauty Expo called BIOMILK Natural Probiotic Skincare offers "potent probiotic protection" in the form of a "probiotic day cream" and "probiotic night cream" that "protect your skin from internal and external attacks." The imagery on the packaging is milk-based, along with pictures of apples and broccoli and other "super foods." The messaging stands out for the sense of fostering the microbial ecosystem rather than cleaning it off. The founder, Valerie Casagrande, formerly worked in sales at Johnson and Johnson. She told me she started BIOMILK when she realized that "probiotics are not just a trend, like coconut water a few years back. This is really going to turn the industry on its head."

Other companies use the term "prebiotic"—meaning a compound intended to "feed" or otherwise foster growth of microbial populations, though the product isn't a microbe itself. This is an even more ethereal claim, since no one knows exactly what products would help feed a person's skin biome (other than one's own sebum). Though many ideas are plausible. I spoke to Stacia Guzzo, the founder of a deodorant brand called SmartyPits, who claims that it can reshape the armpit microbiome. Technically all deodorants do, but the idea of marketing products accordingly—feeding or reshaping the bacterial populations, instead of simply annihilating them—is sound and ahead of its time.

For all the hope and energy in the place, it's unlikely that any indie seller will deliver *the* product that brings skin probiotics into mainstream consciousness. Doing so requires a conceptual shift that historically has happened by way of tremendous marketing and advertising efforts, of the sort only at the disposal of multinational corporations. When the big pharmaceutical industry and soap corporations decide to move into the skin-probiotic space, though, these products could be alongside soap and shampoo and conditioner and lotion and deodorant, on every bathroom shelf.

And it turns out that machinery is already in motion.

• • •

I blame my "obsession" with the skin microbiome on the person who first got me interested in minimalist regimens: Julia Scott, a science journalist based in the Bay Area. Scott had written a fascinating story for the *New York Times* in 2014 about a company called AOBiome that was selling bacteria in a spray bottle for your skin. This really put the company on the conceptual map. I visited her apartment to talk about the story and found a shower that looked strangely barren. She had only a bit of soap for occasional use, and otherwise no products.

The year before, renowned New York University microbiologist Maria Dominguez-Bello and colleagues had published findings that the remote Yanomami tribe in rural Venezuela had the greatest microbial diversity ever discovered in humans. Like the Amish allergy study, it furthered the narrative that the nature-withdrawn, post–Industrial Revolution lifestyle had changed our guts and skin.

The idea was quickly commercialized. MIT-educated chemical engineer David Whitlock, who famously claims not to have showered in more than fifteen years, along with partners, created AOBiome, a company aimed at changing the way we think about bacteria on our bodies. Its products are predicated on a return to nature. AOBiome's first probiotic spray, sold over the counter as a line called Mother Dirt, was intended to help a bacterium called *Nitrosomonas eutropha* recolonize our skin. The pitch is that these ammonia-oxidizing bacteria used to be part of our skin microbiome, where they helped break down smelly by-products of other bacterial reactions that cause odor. But thanks to all the surfactants we use to clean our skin, and our physical separation from the dirt in which these bacteria come to us, the *Nitrosomonas* have all but disappeared.

The claim is that reestablishing them on the body promotes skin health and reduces the occurrence of skin pathologies such as acne. "Within two weeks of use, the AO+ Mist improves the appearance of skin issues including sensitivity, blotchiness, roughness, oiliness, dryness, and odor by replacing essential bacteria lost by modern hygiene & lifestyles," goes the marketing for Mother Dirt's flagship product. Whitlock uses the stuff. Because of it, he says, he doesn't need to shower. (Others insisted to me that, in fact, he needs to.)

I visited AOBiome's San Francisco lab in 2015, while working on a story for *The Atlantic*, and one of their scientists, Larry Weiss, sprayed the bacteria on my face. He asked permission first, but it still felt like being sneezed on. Ultimately I noticed no change, for better or worse. But it did get me thinking about the skin microbiome, and how I should be cultivating it—or not messing it up with my random application of whatever cleaning products I had because I heard about them from a friend or on a podcast or because they were the most attractively labeled at the drugstore.

After I stopped showering and started working on this book, I visited the AOBiome headquarters in Cambridge, Massachusetts. AOBiome now describes itself as a "clinical-stage microbiome company" focused on "therapeutics for inflammatory conditions, central nervous system disorders and other systemic diseases." In its office I meet Whitlock himself. I shake his hand and am not offended, despite his having spent the better part of two decades without a shower. For a place developing bacterial products meant to return our skin to premodern times, the vibe is incongruously start-up, with standing desks and a dead silent open office plan and a half-eaten birthday cake on the counter in the communal kitchen.

That may be because it's really a pharmaceutical company. The company currently has six clinical-stage programs, which include testing bacterial sprays to treat acne, eczema, rosacea, and allergic rhinitis—as well as earlier-stage programs targeting gut and pulmonary disorders. Their new CEO, Todd Krueger, came from a business development background. He has an MBA from Northwestern University, spent time at Bain and Company, and eventually worked on the commercial strategic development of genomics products. Krueger shows me around the tech incubator, and we have lunch at the MIT-adjacent Café ArtScience.

"I think people are probably not going to give up showering," he says, eating French fries and looking at me slightly askance, "and we're not advocating anybody should give up showering. We do think that showering with chemicals is probably not your best solution. Anything with a preservative in it is probably damaging some part of your microbiome."

Soap?

"Well, soap is bad, too. True soaps are really bad."

This was the sales pitch, reader, but I wanted to hear it play out.

"Frankly speaking, most of the bacteria that we get comes out of animal shit," he says, referring not to his product but to humans generally. "When you're born, I don't know if you've studied this, but the bacteria you get from your mom is not from the birth canal, it's from the bacteria that's around your ass, basically."

Many species are common to the vaginal and gut biomes, and the degree to which they each contribute to populating an infant is unclear. But both are known to change during pregnancy, and it is clear that they serve as a sort of inoculation right at birth. The presence of *Staphylococcus* in a mothering vagina, for example, has been found to correlate with infants' likelihood of having asthma at age five. Evacuating one's bowels is common during vaginal delivery. Studies have found that infants born by C-section have a less diverse microbiome than those who had a vaginal birth. When women receive antibiotics during pregnancy, the biomes of their infants are also likely to be less diverse than had they not. The practical implications of all this remain to be seen—of course, C-sections are sometimes vital, lifesaving interventions. But the proper way to expose the infant to microbes after a C-section remains to be studied. For now, multiple scientists I spoke with are in favor of taking swabs of the maternal vaginal microbes and brushing them onto the infant's skin.

This may be the most "natural" approach to populating a child's microbiome. But thereafter, the question of how to maintain healthy exposures is where companies like AOBiome see an opening. As Krueger pitches it, "You may declare war on your microbiome in the shower every morning, and then spray it back on."

A new, daily-use hygiene product is the holy grail of this industry, and probiotic products stand to become that. This helps explain why venture-capital funding has poured into AOBiome and its ilk. And a single probiotic product is small potatoes compared to Krueger's grander vision. "We're only spraying one bacteria; it doesn't mean there shouldn't be thousands and hundreds of thousands of bacteria sprayed back on," he says. "I just don't know what all those things are yet."

The main barrier is that people don't know they want or need to spray themselves with bacteria. Krueger explains to me the difference between primary and secondary demand: Primary demand is when you decide you need a car. Secondary demand is when you're convinced you should buy a Ford. Generating primary demand requires a paradigm shift, and that seems to be what has kept the skin probiotic market from exploding. Once the paradigm has shifted—once people are generally interested in cultivating skin microbes as opposed to cleaning them off—it's much less difficult to get people to choose your product from among various options. That just means flooding their feeds with your brand name.

This shift is underway. A few months after I met with Krueger, Bloomberg reported that AOBiome had licensed its consumer products line to a shell company of S. C. Johnson & Son Inc., the household goods giant that sells cleaning products from Windex to Mrs. Meyer's hand soap. Unilever and the Clorox Company have also made initial investments in probiotic brands—a momentous direction for empires built on removing germs.

The dawning understanding of the microbiome is even starting to influence the marketing of long-standing product lines. In the fall of 2019, Dove launched a campaign on its website giving tips for "caring for baby skin's microbiome." It urges parents to care for the microbiome that "helps keep baby skin healthy by protecting it from harmful bacteria and generating important nutrients, enzymes, and lipids for the baby skin's function." Parents are urged to wash their baby with Baby Dove Tip to Toe Wash because it contains "prebiotic moisture."

The product is, like many other washes for babies, primarily water and glycerin. The claim that it is "prebiotic" is based on the idea that any wash that removes less of the skin's oils than other soaps will be minimally disruptive to the microbiome. This is a thin tightrope to walk—selling soap while implying that soap is bad. As in the sale of all gentle and nondrying formulations, the companies grow another step closer to selling nothing at all. But if they can pull it off, Dove and other juggernaut soap brands stand to endure. Depending on the strength and timing of their pivot from antibacterial to probacterial, fortunes will be made and lost.

With so much at stake, I couldn't help feeling that what started out as a fun piece about showering had somehow turned into an inquiry about the future of billion-dollar industries. Some of the most cutting-edge research is coming from people funded by or working directly for companies that are developing products to sell. There are few experts one can talk to who don't have money in the skin care game.

• • •

The headquarters of the National Institutes of Health is like a college campus for scientists and doctors, a collection of labs sprawled across rolling green hills, encircling a world-renowned hospital where the world's most complex medical mysteries come to be solved.

It's one of the handful of pleasantly temperate and not-humid days in Bethesda, Maryland, and I am here to meet the woman whose work first mapped the skin microbiome, Julie Segre. In a 2012 journal article, she deemed the microbiome "our second genome," calling attention to the fact that the microbes in and on us are "a source of genetic diversity, a modifier of disease, an essential component of immunity, and a functional entity that influences metabolism and modulates drug interactions." While many researchers have focused on the gut microbiome, she believes skin microbes haven't gotten enough attention.

She shows me around the campus, and escorts me to where most of the skin research happens. There is additional security here because the facility houses nonhuman primates, which people sometimes try to liberate. In her office overlooking it all, she pulls up a colorful map of the skin microbiome. It looks like a map of chakras or acupuncture meridians. Composed by Segre and her collaborator, Elizabeth Grice, the rendering of the skin microbiome is like a map of the world from several hundred years ago—involving best estimates based on limited knowledge. She likens the present moment to having just discovered a new organ and only beginning to understand it. Anatomists have known since antiquity that we have livers, for example, yet still don't understand everything about how they work (another building is devoted to this). But, she explains, the microbial map is a good place to start.

There are around a billion bacteria per square centimeter of skin. In total there are trillions, across at least a few hundred different species. They vary based on the type of skin environments we all carry around—which break down into three conventional categorizations: oily, moist, and dry. Oily sites are your forehead and chest; moist are the armpits and creases in your elbows and knees and groin ("inguinal crease"); dry is your forearm. The microbiome in the bend of my left elbow is, for example, most similar not to the biome on my left forearm but to the biome in the bend of my right elbow.

These are saline, sweaty environments that will harbor the same microbes reliably, even when cleaned off. It's that environment that makes these regions subject to odor-producing bacteria that simply don't colonize the forearm or stomach—which demand less or no washing as a result.

What we have in our moist creases—Julie's term, not mine—is totally different from what's on our chests. The highest bacterial biomass on the skin surface is in the armpit, where colonies are directly sustained by food resources primarily provided by apocrine glands. In all, though, the skin does not provide very many nutrients. "It's not like the gut where there's just constant food for the bacteria," Segre explains.

The most common bacterial genera on our skin include *Staphy-*

lococcus, *Corynebacterium*, *Propionibacterium*, *Micrococcus*, *Brevibacterium*, and *Streptococcus*. On the map Julie pulls up, the oily sites have a lot of *Propionibacterium acnes*, which does correlate with where acne occurs—hence the name—though the causal relationship is unclear at best. Eczema tends to occur in the flexural creases, like the bend in the elbow and behind the knees. Flares tend to correlate with increases in *Staphylococcus*.

"These diseases are clearly at least related to microbial imbalance," Segre explains. After decades of trying to pin these diseases on an invasive species—an infection in the traditional sense that could be eradicated with antibiotics—it's turning out that the real problem is an imbalance. We've only recently been able to understand this because the technology required to sequence the DNA of all these microbes wasn't available.

Segre is in the first generation of scientists native to this capability, having trained at the Whitehead/MIT Center for Genome Research. She got into the field because, she tells me, "I really like organizing large data sets," and that's what genomics is. Once the world of microbes on the skin started to become apparent, the skin just happened to produce a lot of data sets. She was introduced to skin biology during her postdoctoral research at the University of Chicago with Elaine Fuchs, who won the National Medal of Science in 2008 for her work on skin stem cells. Fuchs's research continued the work of her own postdoctoral adviser at MIT, Howard Green, who had figured out how to grow human skin in 1975. By taking just a tiny, two-millimeter punch biopsy and isolating the stem cells in our skin—the cells that allow us to replace the skin cells we're constantly sloughing and exfoliating off—researchers can culture them to grow all the layers of skin.

Growing sheets of a person's own skin in a lab isn't a purely academic exercise. It holds promise for treatment of burn injuries where people require skin transplantation, since using a person's own tissue dramatically lowers the odds of rejection by the immune system. Lab-grown skin could also be an ideal model for testing the effects of skin products and medicines, as well as microbes. Segre's lab already does some testing this way. Her team gets most of its stem cells via donated skin from circumcisions, she says, but people also donate skin after various reduction surgeries. (It's possible to buy actual human skin stem cells online—a company called ProtoCell, for one, sells a vial of 500,000 foreskin-derived fibroblasts for $489—but these are of zero use unless you know how to grow them into skin.)

Segre does, and using this lab-grown and 3-D printed skin, her team can add microbial gardens and study their functions. Her team will test hypothetical combinations in what she calls "microbe-microbe competitions"—to see how different species interact with one another, and with the skin. Given the number of microbes and the variability in skin, this is the sort of competition that would involve millions of rounds. It's also very expensive. Funding people who truly want to understand the ecosystems, and aren't looking to understand just enough to sell a product is a massive investment. While scrolling through images on her computer, Julie shows me a picture of her with President Obama and Jack Gilbert and some other scientists. Obama had invited them to the White House a few years ago because he wanted to learn about the microbiome. It's a reminder of the importance of government investment in science. Without that sort of public commitment, I, too, would be getting all my information for this book from industries and industry-funded work.

For as exciting as the skin microbiome is to her, Segre is puzzled and almost parentally defensive that it hasn't captured the public's attention yet. "I don't understand exactly why it is that people have such a different sense of the microbes that live in their gut than they do about the microbes that live on their skin," she says. "Everyone wants to eat Activia yogurt and colonize themselves with bacteria, and then they want to use Purell."

The promise she sees at the moment is not in probiotics (which, technically, are the microbes themselves), but prebiotics—the various products that "feed our microbial gardens." The normal and beneficial microbes are there already, in most people; we probably don't need to add them so much as promote them. Many people I talked to who use probiotics conceptualize them as the antithesis of antibiotics. But the inverse of an antibiotic is really the prebiotic: an antibiotic is suppressing something that's in the microbial community, and a prebiotic is promoting something in the community. A probiotic is a whole different concept, actually coming in with an external organism that maybe isn't native to the host.

Understanding, testing, and selling prebiotics may be more straightforward. Lots of things that are already on the market, like the aforementioned clay-based deodorants, are probably working as prebiotics. Another example that's on the market already is ceramides. These are lipid molecules that both naturally occur on our skin, as part of the barrier and lubricating function, and are increasingly sold in skin care products. They may serve as a sort of food for microbes and, in turn, microbes may signal the skin to produce more ceramides. At least, both claims are already being made on products. More research could help us understand exactly what these compounds do to the microbial populations on our skin, and who benefits from putting them where.

"All the ingredients that go into these creams are something that

I think could be prebiotics," says Segre. "It would be interesting to know, Does a given microbe really use this as a carbon source—as a nutrient that they are requiring to grow? I think people are doing these experiments themselves. They're saying, 'I like this cream. I don't like this one.'"

I ask the obligatory question about her personal hygiene, and she says she always encourages everyone to wash their hands with soap and water, and not to take its value for granted. The value of the practice increases especially during outbreaks, like flu or cholera, when any given wash may be the one that saves a life. "On the other side, we're probably overusing—certainly overusing antibacterial soaps, and potentially breaking down the skin barrier by drying ourselves out by using so much soap, contributing to that eczema inflammatory process."

The majority of kids with eczema grow out of it by adulthood. But, Segre says, "If you're just thinking about it like, 'Well, your kids can be miserable, but they'll grow out of it,' what about if I told you that [the eczema is] going to affect your kid's life over their life span? Then you're really, I hope, going to be motivated." Here she refers to the atopic march, where food allergies, eczema, and other immune sensitivities appear together.

Stopping or reversing this is the ultimate goal. Maximizing exposures to microbes, especially early in life, has shown some promise as a prevention. Exposure to skin microbes does affect allergies: a 2017 study by Tiffany Scharschmidt at the University of California, San Francisco, showed that mice who were exposed to certain strains of *Staph. epidermidis* in the first week of life had regulatory T cells that were able recognize it later when she reexposed the mice to the same bacteria. If the mouse had not been exposed before, the bacteria initiated an allergic response.

As with training the immune system to recognize peanuts, the first years of life seem to be crucial. While the immune system remains malleable—and can be influenced by microbial exposures throughout life—in the very beginning it is like freshly poured concrete. After that, we pick up microbes and we lose them, but the foundation stays the same. Permanently changing the basic skin microbiome of an adult seems much more difficult. Segre describes the process: it would mean first making you as germ-free as humanly possible—by bathing you in an agent called chlorhexidine, as is done in hospital ICUs when a patient is extremely sick and has an immune system that can't fight off even the simplest of disease-causing bacteria—and then transplanting a functional microbiome from there.

This has been successfully done with gut microbiomes. Though the skin microbiome involves fewer microbes, the nature of the skin's physiology—and the fact that the skin technically has several different biomes in several different areas of the body—creates new challenges. Immunologist Susan Wong at the New York State Department of Health has studied the effects of the process Segre described, and it seems that since the microbiome is coming from deep in your pores, even that dramatic a treatment will have only a transient effect on your skin. Once a person recovers and is out of the hospital, their skin tends to repopulate with the microbiome that was established in infancy and early childhood.

This makes it unlikely that bacterial sprays will be effective ways of treating people later in life, though there is potential at young ages. Although, Segre says, "There are questions we need to answer before I'm ready to put a live microbe onto a kid." With pharmaceuticals, it's not difficult to calculate the time it takes for a drug to pass through the body. This makes dosing and side effects somewhat predictable. It's also guaranteed that eventually your body will clear the drug. "With a live organism, it's not even a given that your body will clear it."

Others are ready to test the possibilities. In 2018, headlines declared the first successful use of a probiotic treatment for eczema, which was long thought to be due to an overpopulation of *Staph. aureus*. Indeed, inflammatory proteins coming from this bacterium do seem to cause the infamous itch that sets off a flare and is exacerbated by subsequent scratching. Instead of attempting to eradicate that bacterium, researchers at the National Institute of Allergy and Infectious Diseases sprayed patients with a different bacterium. After applying *Roseomonas mucosa* on their inner elbows twice weekly for six weeks, most of the patients saw improvements in symptoms—less redness and itching, according to lead researcher Ian Myles. Some people reported needing fewer topical steroids even after this "bacteriotherapy" stopped. Myles's team then repeated the experiment in kids and found the same results—in addition to a decrease in the amount of *Staph. aureus* on their skin.

"By applying bacteria from a healthy source to the skin of people with atopic dermatitis, we aim to alter the skin microbiome in a way that will relieve symptoms and free people from the burden of constant treatment," Myles said at the time. He added that if future clinical studies show that the strategy is effective, longer-lasting changes to the microbiome could avoid the need for daily application of products.

Though altering the skin microbiome is a new conceptual approach, it's also likely what we've been doing indirectly for a long time. Once a person has eczema, the current standard treatments are antibiotics, steroids, and emollients (moisturizers, creams, or lotions that mimic the oils secreted normally by the skin). Segre believes emollients may work not simply by restoring the skin barrier, but by feeding other microbes—promoting the growth of, for example, *Roseomonas* or *Corynebacterium*, which were somehow not getting enough resources and were being outcompeted by the staph. But the aforementioned treatments take time to work, if they do at all. Many people have to apply them several times a day.

Breaking the cycle of flare-ups, steroids, and antibiotics is where, hypothetically, there's room for a probiotic—something to actively and immediately repopulate the skin. This has been tried with local transplants from a kid's own nonflared skin. The challenge is understanding the specific microbial community and shifts associated with eczema.

Could we figure out when a kid is potentially going to flare, and initiate treatment in anticipation? In a dream scenario, you could regularly test your skin to predict these debilitating flares before they occur and start a vicious cycle.

Richard Gallo at University of California, San Diego, has also partially transplanted skin microbiomes from one part of a kid's skin to another and had some success. In a paper published in February 2017 in *Science Translational Medicine*, his team reported isolating and growing "good bacteria"—*Staph. hominis* and *Staph. epidermidis*—that produce antimicrobial peptides to defend against *Staph. aureus*. The researchers isolated the compounds, formulated them in a skin cream, and applied them (or "transplanted" them) to the forearm of people with eczema—and they saw improvement in symptoms.

"We discovered that healthy people have many bacteria producing previously undiscovered antimicrobial peptides, but when you look at the skin of people with atopic dermatitis, their bacteria are not doing the same thing," Gallo said at the time. By his interpretation, for all the work that's gone into antibiotic development, the chemicals produced by normal skin bacteria may be the best tool to fight an imbalance of skin microbes. The lead researcher on the project in Gallo's lab, Teruaki Nakatsuji, called this transplant a "natural antibiotic" and suggested that it would be a way around killing innocent-bystander bacteria and contributing to antibiotic overuse and resistance.

Among other "natural antibiotics" that seem to help moderate the growth of *Staph. aureus* during eczema flares is, simply, sunlight. Researchers in Norway found that when people were exposed to UV-B rays regularly for four weeks, their colonies came back toward the norm.

In addition to all the ways skin microbes could inform new treatments—or simply explain how old ones work—Segre believes their most promising use may be as a predictive tool. Different people respond to different things, and it can be difficult to predict what will work for whom. Every forum where anyone is raving about a skin-care product will also include people who swear it's a waste of time and money.

Treating skin conditions may not have to be such a deeply frustrating and sometimes harmful series of trials and errors. In the next few years, dermatologists could potentially sequence an individual's skin microbiome, then differentiate it from their microbiome during an eczema flare. Even if eczema is not one single disease, that would be a way to see exactly what was causing flares in that particular person's case. It could then be possible to use more targeted approaches to get them back to their normal state. Sometimes that might involve a probiotic or prebiotic instead of an antibiotic.

Meanwhile, scientists like Segre are focusing their limited budgets on urgent threats to humanity. At the top of her mind is a lethal "superfungus" that her department has been scrambling to try to understand. No one even knew the species existed until a decade ago, but it has now emerged as one of the CDC's top concerns. It's called *Candida auris*, and it has Segre "captivated."

In 2009, researchers reported the discovery of a new strain of fungus in the ear canal of a patient in Japan. A few years later, the fungus was linked to mysterious cases of bloodstream infections in hospitals around India, and before long more strains of *Candida auris* were appearing around the world. *Candida auris* colonizes human skin, and can then get into the bloodstream when a nurse inserts an IV. It's popping up right now in skilled-nursing facilities and long-term, acute-care hospitals. *Candida auris* was first detected in the U.S. in 2013, and in April 2019, a major story in the *New York Times* reported that at least 587 cases in the country had since been confirmed. By October of the same year, the total had reached more than 900. The new strains are referred to as a superfungus because they are resistant to the antifungal medications used to treat them.

Every microbiologist I spoke with agreed that antibiotic overuse is likely a bigger contributor to messing up our microbiomes—gut and skin—than hygiene itself. We may not be doing a lot to change our biomes by showering less, but by conceptually getting over the idea that microbes are bad, we might consume less and use less of the antimicrobial products that do indeed create microbial "superbugs" that threaten all nonmicrobial life on Earth. Our concept of cleanliness is, in other words, inordinately consequential.

Segre believes overall it's great that the public is starting to glom on to the idea that "you're a superorganism"—that all these microbes live in and on you and need to be considered in every decision we make about what we consume and apply. As recently as 2013, when researchers published a paper in the *New England Journal of Medicine* about fecal transplants as a promising cure for the often-fatal bowel disease *Clostridium difficile*, many who read the news expressed disgust or indignation that this was even being tested. It seemed so contrary to the staid and sanitary ethos of medicine that even some physicians were dismissive.

Now, as probiotics fill shelves and refrigerators, fecal transplants are fast becoming a part of clinical practice. Though the practice is still in its infancy, and much remains to be understood (including some unexpected effects, such as people who have been lean or obese their whole lives suddenly gaining or losing weight, as if the new biome has altered their basic metabolic set points), some patients have seen lifesaving results. I mused to Segre that it's fascinating that the U.S. health care system is the seventh largest economy in the world, and one of our most exciting breakthroughs in recent years is putting other people's feces into our bodies.

"Anyway," I say, as we both stare into the middle distance, "I should let you go."

"Yeah, it's getting late."

She rides the elevator down with me, and I tell the security guard I don't have any of his marmosets. In parting, I ask if she knows of anyone doing exciting work in the development of microbial products for the skin, without the hype. Segre suggests that I talk to Julia Oh. And so I wander off of the NIH campus and head north, to where state-funded basic science is translated into corporate profit.

• • •

At the Jackson Laboratory for Genomic Medicine in Farmington, Connecticut, Julia Oh is making skin probiotics a reality.

Oh studied fungal chemogenomics at Harvard and then went to Stanford to study the genomics of wine yeast before focusing her attention on skin microbes. Wine is important to people, she reasoned, but skin is even more so. And she is certain that the skin microbiome plays "an active and intimate role in shaping the health of skin."

The challenge is to figure out just how our microbes interact with our skin cells. In 2017, Oh received a New Innovator Award of $2.8 million from the NIH (meant to support "exceptionally creative early-career investigators who propose innovative, high-impact projects") to study how to develop engineered probiotic treatments for a variety of skin and infectious diseases. It's an indicator that the leading power brokers in skin research believe that the next generation of products will involve harnessing and manipulating the powers of the microbes on our skin—as well as those no longer or not yet there.

The premise of her grant is to better understand how probiotic strains integrate into an existing microbial community. Her lab is also using both experimental and computational modeling to understand how the skin microbiome is assembled, how resilient it is to perturbation, and what factors determine how well foreign microbes can compete and integrate into an ecosystem. It sounds to me like the skin care equivalent of landing a person on Mars, but she's confident.

She sees the promise of probiotics to change the way we treat skin diseases: by shaping the immune milieu in ways that make you better able to eradicate skin infections or skin disease, or by reducing unnecessary inflammation.

The challenge in adding microbes to tune and modulate the immune system is in getting specific microbes to stick around. In one experiment Oh's team doused mice with microbes three times a week for twenty weeks, and by the end the new microbes comprised only 2 percent of the mice's overall skin biomes. That told the researchers that, at least for these microbes, there are innate forces that make a skin microbiome more or less resilient to accepting them. Some strains colonize one mouse but not another. Is it simply a matter of space? Are the microbes in the hair follicle basically preventing anyone else from colonizing ("seat's taken") because the ecosystem is already at capacity? Is it a matter of limited resources? Are microbes being turned away by an immune response?

Changing a person's skin microbiome in predictable, safe ways means understanding how all these forces work together. Certain microbes, such as staphylococci, can secrete molecules that prevent colonization by other microbes. Others can trigger the skin's immune system in ways that make life harder for their colleagues. Still other microbes are riding along, slurping up skin oils and then secreting acids to lower the skin's pH. All of which is to say that even adding a seemingly inert microbe to the mix can throw things out of whack in unexpected ways.

While these interactions are being explored, Oh and colleagues are forging ahead with another approach: Instead of trying to change the populations, they can use the microbes that are already on us to help deliver medications to our skin. The researchers think of the microbes as a chassis for different therapeutic agents that might alter our immune responses.

That starts by studying which microbes can activate which types of immune cells. Oh is cataloging the known interactions between skin microbes and the immune system in the skin. The idea is that this catalog could be compared with any given individual's microbial and genetic map, and, theoretically, used to identify the cause of whatever symptoms that individual is experiencing.

Oh has also developed CRISPR-based gene-editing tools to help figure out not just which bacteria are doing what, but which characteristics of that bacterium are actually responsible for changing the immune system. Oh has been collaborating with biochemist-turned-venture-capitalist Travis Whitfill to apply this idea to the treatment of skin disease. Whitfill is a cofounder of Azitra, a biotech company for which he made the *Forbes* 30 Under 30 list in 2018 after raising $4 million "to try to leverage benign bacteria living on the skin in order to treat skin diseases." By the end of 2019, the company had raised over $20 million.

Azitra's goal is to turn bacteria into the world's tiniest drug dealers. Using *Staph. epidermidis*, a species that lives on most skin (and, so, could be easily transferred to a patient), Oh and other researchers have been able to genetically modify the bacterium to secrete various immune-modulating compounds—or, as Whitfill calls them, "assets" that secrete "therapeutically relevant proteins." His hope is that these bacteria could help treat various skin diseases and their symptoms. Azitra is now testing to see if these proteins have the intended effect in real life, and how to get the right doses to people. Whitfill believes the most promising potential for this work is treatment of rare genetic skin diseases where a certain protein is missing. Current treatment options often mean applying a cream multiple times per day, or taking a pill that might have effects on other organs of the body. By comparison, populating the skin with drug-secreting microbes could supply a constant stream of treatment.

For example, infants with Netherton syndrome have skin that is brittle, scaly, and especially porous. It can leak fluid, putting the child at risk of dehydration, and is prone to invasion by microbes that can cause life-threatening bloodstream infections. People who survive into adulthood continue to have flares throughout life, sometimes triggered by stress. Most also have immune-related conditions like food allergies, hay fever, and asthma. The whole cascade seems to start with an overactive enzyme that causes skin to break down. In lab settings, this enzyme can be blocked by a protein called LEKTI. In 2019, Oh and colleagues announced that they were able to make a strain of *Staph. epidermidis* that secretes LEKTI. Theoretically, introducing this bacterium to these patients' skin would help their symptoms. This is currently being tested in a clinical trial.

Another of Azitra's proprietary bacterial strains is aimed at helping to treat eczema. By equipping the bacteria with a gene to make the protein filaggrin, which binds keratin filaments, it could help seal the skin from the outside world. In the absence of filaggrin, breakage of the skin lets in antigens that lead to inflammation. Minimizing the tiny breaks that are associated with eczema flares could, hypothetically, help stop or prevent those unpredictable episodes of extreme itch and redness.

The idea of selling people genetically modified bacteria to put on themselves may not seem immediately ideal as a sales pitch. It's antithetical to most ideas of cleanliness throughout history. But Whitfill believes people will be forced to reconceptualize their skin as treatments like this become more common. While the therapies Azitra is developing would be tested and regulated as prescription drugs, in early 2020 the company also announced a partnership with Bayer to create cosmetic and personal care products containing *Staph. epidermidis* that hasn't been genetically modified. These could come to market much more quickly since they don't have to go through the rigorous testing processes that drugs do. They simply couldn't claim to treat or cure a disease. "You can't say on the label it cures eczema," Whitfill explained to me, "but you can say stuff like 'for eczema-prone skin.'"

Though these products may or may not prove to help people with eczema, marketing claims will imply that they do. Combined with an enormous and solution-hungry consumer base, the stage is set for a major mainstream skin probiotic market, in addition to people who use very similar strains as prescription drugs. Whitfill described the basic concept in a 2018 interview on a marketing podcast aimed at biopharmaceutical investors, and there was a moment when dollar signs presumably glinted in the eyes of the host as he connected the dots: "This opens up for you other markets such as cosmetics and beauty supplies, where costs are different than producing a [pharmaceutical]."

"Exactly. It's not as cheap as making yogurt, but it's not too far off," Whitfill responded. "There are a lot of potential plays there with consumer health and over-the-counter products, things like that." Using a novel FDA pathway called "Live Biotherapeutic Products," this sort of bacteria could be taken almost directly to market.

"People are starting to become aware of the microbiome, and a lot of consumer products are not microbiome friendly," Whitfill continued. "This would be different in that it's compatible with the biome in a natural, safe way, restoring the balance of your microbiome. In doing market research, we feel confident that consumers want this product."

While most people I've talked to aren't extremely excited by the idea of dousing themselves in bacteria, appealing to the sense that a product is "natural" or would "restore balance" would be a time-tested and proven approach to take these products mainstream. Of course, a product that people want to buy is not necessarily the same as one that works and is safe and beneficial to people. And the effects of such products are likely to vary widely from one person to the next. Each of us has unique skin with a unique microbiome and a uniquely calibrated immune system—a result of the cumulative interactions among our microbes, our exposures, and our genetic predispositions. Populating a person with living bacteria that can reproduce and thrive—or die off—makes it much more difficult to ensure that dosing is standardized.

Given all the variability among people, and the difficulty of populating skin with new microbes for any significant period of time, the optimistic premise quickly seems radically complicated. In trying to make sense of what can actually be done to help anyone currently—as development of these theoretical treatments plays out—I'm drawn back to the biodiversity hypothesis. Applying genetically enhanced skin microbes after symptoms appear is one approach, but another is to try to develop healthy biomes in the first place. The time-tested way to optimize the immune system still seems to be through once-common, diverse, early-life exposures. As Oh herself concludes, "If your skin biomes are diseased, then we have to have therapeutic targets. But if you already have a healthy skin microbiome, it's a more do-no-harm philosophy."

IX

REFRESH

Lying in a hotel bed one night in 2008, a corporate salesperson named Shawn Seipler was contemplating the meaninglessness of existence.

He worried he wasn't doing anything productive for the world, arranging tech-industry partnerships. He started considering the wastefulness of his travel schedule. Not just the carbon footprint of flying all over the country—staying in a different hotel many nights of the year—but the little things. He called down to the front desk of the hotel and asked what they did with all the *soap* that was left behind.

Unsurprisingly, the concierge said they throw it away. Seipler thought about the scale of that waste, multiplied by the number of hotel visits in the U.S. alone, and estimated that every day we're collectively discarding around five million bars of soap.

This did not help him fall asleep.

American hotel operators adopted the European concept of offering soap to guests in the early 1970s, as ever budget-conscious chain hotels fought for ways to distinguish themselves from competitors without significant investment. A little bar of soap in the bathroom or a mint on the pillow made you feel cared about, even if the environment was otherwise dreary (and, if you took a black light to it, unclean).

Including a previously used bar of soap, or a half-empty economy bottle of shampoo, did not apparently confer the same feeling of care as giving each guest a new one. Each item needed to be individually packaged, wrapped in pleats of decorative paper or poured into tiny plastic bottles, and discarded after each guest departed even if not obviously touched.

For a person with a tendency to be conscious of waste and their own imprint on the environment, this could trigger a panic attack. (Some people are now being diagnosed with "ecoanxiety.")

Seipler channeled this energy productively. Newly imbued with purpose, he went on to found an organization that recycles used hotel soap, melting down the partially used bars and fashioning them into new ones, putting them in fresh wrappers (as consumers have come to expect), and distributing the new bars of soap to people in need. The group is called Clean the World.

At their Florida headquarters, huge bins of partially used soap are separated by type—so as not to mix colors and scents when melting them down and reshaping them. The bars are shredded and melted together; the heat from the process sanitizes any lingering human matter, and the end product is a good-looking bar that can be sent around the world. The group says they have sent some fifty million bars to more than one hundred countries.

The effort became part of a story that's much bigger than wasted soap. While Clean the World is mainly concerned with preventing hygiene-related infectious diseases, it's also about the dignity of having basic needs met. The overuse of hygiene products is more than an environmental problem—more than antibiotics and plastic shampoo bottles accumulating in trash islands in the ocean. Even the aforementioned problems of autoimmune diseases, of eczema and acne and asthma and all else that we may have unleashed on ourselves in wealthy countries, are only part of the issue.

At the same time as the conditions of excess and isolation are on the rise in much of the world, elsewhere preventable death and infectious disease plague the two billion people who lack access to basic sanitation. In 2019, UNICEF reported that one third of the global population does not have regular access to clean drinking water, and even more do not have a way to wash their hands at home with soap and water. The world's hygiene problem is one of not simply too much or too little, but of radical imbalance.

Clean the World is part of the United Nations Water, Sanitation, and Hygiene (WASH) program, an arm of the ongoing effort to "end extreme poverty, reduce inequalities, and tackle climate change globally." Much of this comes down to soap and water. The most globally visible problems involve disaster situations. For example, after the earthquake in Haiti in 2010, some 8,000 people died of cholera—which is entirely preventable with clean water and hygiene.

But much of the effect of poverty on health involves day-to-day hygiene habits. In 2020, the leading global cause of death for children under five is still hygiene-related illnesses—primarily diarrhea and pneumonia. Around 90 percent of these deaths are estimated to be preventable with hygiene, sanitation, and clean water. In terms of lives saved per dollar spent, there may be no more effective medical investment than universal access to hand-washing facilities and toilets that drain far from sources of drinking water.

One of the countries where the problem is most evident is

Mozambique, where, according to the UN, half the population lacks access to a safe water supply. Meldina Jalane, a forty-five-year-old housekeeper, grew up in the country's most populous city, Maputo. She spent her childhood and early adulthood carrying water to supply her family's home, she tells me matter-of-factly. She and her siblings would trek to the borehole four times a week—usually at night, to avoid the heat and increase the odds that the well would not be dry—and carry water back to fill the home's water tank. When she was ten, she carried five liters. By adulthood she was a professional water carrier, bringing forty liters at a time (on her head) to construction sites to mix cement.

Even when her family's water tank was full, there was still a step between pouring and drinking. All the water had to be boiled to be safe. Jalane now uses a product called Certeza, which can be put in the tank to purify it. The product, which translates from Portuguese as "certainty," is a point-of-use water-diluted sodium hypochlorite solution. It was launched in 2004 and sold at subsidized prices through the private sector. It is an individualistic approach to clean water—rather than pooling resources and sanitizing water centrally, each person carries around little bottles of a product that can be poured into water before drinking.

These are distributed by a mix of government investment and international aid. Brenda and Stephen Valdes-Robles, Americans who worked with USAID on this and other projects around Maputo over much of the past decade, describe the scope of the problem as one that's inconceivable in much of the developed world. For most of history and still in much of the world, a reliable public water supply—of the sort many of us now take for granted—has been synonymous with luxury.

Jalane recently visited the U.S. for the first time. In New York, she said what fascinated her most about the city was the regularity with which the trash is collected. Within the U.S., the city has something of a reputation for smelling like garbage, especially in the summer. It really is miraculous that the New York Department of Sanitation is able to keep a city this dense ever not smelling like fetid waste.

Despite the preciousness of water, and perhaps because of it, appearing clean has always been a priority for Jalane and her family—worth walking through the night for. If there was any hope for the upward mobility she ended up achieving, by the social standards that associate cleanliness and status, she couldn't afford to look any other way.

• • •

The situation in Mozambique is far from unique. We are—and always have been—in the midst of what public-health experts deem a "global water crisis." For all the lofty innovations touted in Silicon Valley and at global health tech conferences about curing rare metabolic syndromes and unwinding complex cancer pathophysiology, the medical community is still struggling to address a seemingly simple need: getting people water and toilets.

"Hygiene is one of the most cost-effective health interventions on the planet," says Sarina Prabasi. "And that's just soap and water, mostly." When we spoke, Prabasi was the CEO of WaterAid, a global nonprofit that sees health and water access as one and the same. WaterAid works in countries with the least access to water, building rainwater tanks, hand pumps, and wells. Her group estimates that 60 percent of the world's population lives in "water stress," meaning they don't have clean water close to home that's safe to drink.

Not only does this mean that hundreds of thousands of children die of infectious diseases, it also means that producing water becomes a top concern and time-consuming activity for much of the world's population. Prabasi was formerly based in Ethiopia, where she helped tackle trachoma—the leading cause of preventable blindness in the world. "It was one of the most horrendous examples of suffering I've seen," she says, describing the debilitating infection in which the eyelashes turn inward. "It's mainly related to hygiene. This can be prevented with face washing."

Common among preschool-age children, the infection can be cleared by the immune system. But after years of repeated infections, the inside of the eyelid scars down, and the then-inward-facing lashes scratch the cornea to the point of blindness. This happens up to four times more often in women than in men, likely due to their disproportionate responsibility for child care. The World Health Organization put the annual economic cost at around $8 billion.

While diseases like trachoma are now localized issues, Prabasi sees much more universal challenges in what's often the most difficult subject to address: menstrual hygiene. The effects of ignoring this basic element of human biology resonate across the global economy, affecting the position of women in every country that does not provide comprehensive education and readily accessible hygiene products (including the United States). Around the world, millions of girls drop out of school each year because of hygiene issues relating to menstruation.

"Menstrual hygiene has become a much bigger part of our work over the years," says Prabasi. She noted that stigma over menstruation is especially strong in Nepal, where it's not uncommon for young women to avoid going to school for four or five days a month because they don't have a way to manage menstruation.

"In many cases there's no privacy or bathroom in the school," she says, "which means they fall behind academically, which then means they're much more likely to drop out of school altogether."

In Mozambique, only 25 percent of rural schools have restrooms.

In such cases of severe disparities in health due to hygiene, even very small interventions—maybe even the cost of a single jar of non-GMO probiotic facial cream—could completely change a life's trajectory.

. . .

While writing this book I was in a meeting with the dean of the Yale School of Public Health, where I teach, discussing what felt like the only thing I'd been discussing for as long as I could remember—hygiene and skin microbes—when his eyes lit up. He explained that, decades ago, in response to a spate of deaths, he had done several studies of douching.

The fallout from douching—the act of flushing the vagina with water and other products in order to "clean" it—may actually be the first widely recognized instance of the negative effects of hygiene on the microbiome. Though public-health campaigns have slowly dispelled ideas about the necessity or safety of douching, it was widely practiced for centuries. In the 1940s, Lysol was advertised as a way to "safeguard feminine allure" with its "amazing, proved power to kill germ-life on contact," which "truly cleanses the vaginal canal." Many doctors either recommended it or thought it a harmless matter of grooming, until epidemiologists started reporting higher rates of infections among women who douched.

It was difficult to convince the public that disinfecting or rinsing anything could be harmful—unless the products being used were contaminated. But the infections weren't the sort that could come in a bag. They were often gonorrhea and chlamydia. My dean, Sten Vermund, reasoned in a 2002 paper coauthored with his colleague Jenny Martino that cleansing must be depleting the normal microbes that are supposed to live in the vaginal canal. With those gone, the tissues were open to inhabitation by sexually transmitted infections that filled an "ecological niche."

In Alabama, where the two were working at the time, doctors were seeing cases of life-threatening peritoneal infections and ectopic pregnancies as a result of douching, especially among African American and Hispanic women. The infections spread up the reproductive tract and throughout the pelvis. Though risks had been reported in some places for a while, populations with less access to medical care tended to be at the highest risk of misinformation and targeted marketing, and of advanced infections due to lack of primary care.

Part of why it took so long to figure out what was going on—and why douching is still practiced by some women in the U.S. and around the world—was a lack of conversations around women's health. A similar lag in public response was notoriously part of the story of toxic shock syndrome, the deadly condition that happens when the immune system goes into overdrive in response to *Staph. aureus* growing on tampons—often related to leaving them in too long. Many serious illnesses and deaths could have been prevented with better information, willingness of everyone to talk more about all manner of hygiene, and, of course, access to tampons. They are often left in purely as a matter of limited resources. Tampons and other menstrual products are some of the only hygiene products that could be deemed essential. Yet in most U.S. states, they are actively taxed. This is despite a federal law that prohibits taxation of medically necessary products.

The taboos around sexuality and hygiene that combined to keep these problems understudied and underdiscussed also extend to anal hygiene. The fact is that most infectious disease comes down to waste management, and washing your hands after using the bathroom is only really about removing fecal matter. Most people don't wash their hands thoroughly or at all after defecating, even in places where soap and water are right there. Researchers at Michigan State studied the rate of thorough hand washing in public bathrooms in 2013 and put it at 5 percent.

The hands, of course, aren't the only place that needs cleaning. Visitors from wealthy countries are often horrified by the American standard of toilet-side practice. The marketplace has been dominated by dry toilet paper for decades. There are few ads for "a better way to clean your ass." Prepackaged wet wipes have seen a surge in popularity recently, though they can pose environmental issues, namely clogging sewers. They also simply cost a lot. Newer brands make biodegradable products that claim to be flushable, but they remain much more costly than rolled paper. The environmental costs of transport add up, too. The extremely reasonable, hands-free solution that much of the world has discovered, the bidet, remains out of the realm of discussion for many Americans.

When people hear you stopped showering, you can tell almost everyone is thinking about toilet-related hygiene—but only some people actually ask about it. The dearth of bidets in the United

States is even a reason that some people shower—because dry toilet paper is inadequate. You wouldn't come in from gardening and wash your hands with a dry paper towel, so why would dry paper be the standard for cleaning off actual fecal matter?

For all the prosperity and invention of the "greatest country in the world" or the "shining city upon a hill," we have made no progress in the domain of cleaning our anuses. The great forces of American capitalism have left the marketplace largely untouched. Even the Romans had contraptions that outperformed dry toilet paper, preferring sponges on sticks.

So this is a topic I feel I must address. If you don't like the word "anus," you may want to skip this bit and go for a long walk and reflect on that fear of anuses. Say the word aloud, again and again, louder and louder, until it loses its power over you. This is the first step to cleaning the anus effectively and efficiently, which is a step toward major environmental and health benefits on a global scale. The pejorative associations are unfortunate, because it's a remarkable body part.

I haven't invested in a mechanized Japanese toilet (though people swear by them) or even a simple bidet, nor have I fashioned a sponge on a stick. But I do have a great toilet paper hack.

The secret is that extremely adequate anal hygiene can be achieved in the same way as so many other things in life, by using water and a little bit of toilet paper. Hold it under the sink to wet it, and then wipe. That's it.

The follow-up question tends to be, "Won't it fall apart when wet?" The answer is no, not unless you're trying to drown it. A small amount of moisture can do more than could ever be accomplished with dry paper, and with less need to purchase expensive ultrasoft toilet papers (some of which advertise that they contain *moisturizers*). This means using less-expensive paper and less of it, since the water makes the process so much more efficient.

Public restrooms do pose a challenge. It is uncouth to stand near the sink to do one's wiping. That is, unless you're doing so to make a point and raise awareness that across the world, nearly seven hundred million people have to defecate in the open. In Madagascar, Mozambique, Namibia, and Zimbabwe, the proportion of rural people that practice "open defecation" is higher than the proportion that have access to a basic latrine.

Either way, just make sure to wash your hands.

. . .

The skin ecologist Jenni Lehtimäki was walking through a residential neighborhood in Copenhagen in 2017 when she came upon a playground full of kids. "I was so surprised!" she recalls with genuine delight, because unlike most playgrounds, this one had cows. Looking around, she saw that there were also chickens and goats and small ponies. She recounts the scene with the expected enthusiasm of an ecologist stumbling upon an ecologically diverse urban habitat: "I was like, *What is this place?*"

The place wasn't a zoo, but a simple public park that also had animals, called the Bondegården ("farm"). It turns out to be one of several where the Danish government promises parents, "Your child can experience, touch, and see lots of different animals." It also offers after-school programs where children actually do work to take care of the animals. Lehtimäki loves this not just because it teaches children the responsibility and grit born of corralling chickens, but because it also exposes the kids to some nonhuman microbes, which she believes could have some real benefits. Lehtimäki says this is exactly what she'd like to see back home in Finland. (To hear that even people in Finland envy other countries' social progress is cathartic.)

This nature-based approach to child care is indeed spreading. Finland does have some nature-oriented day cares—and they are not just about drawing trees and having the children read Emerson. "They spend their whole day outdoors," says Lehtimäki. "Even when it's winter—it can be minus twenty-five [Celsius] and so cold." Parents are advised to dress their children in many layers, according to Finnish news coverage of one such day care. It also suggests that the three- to five-year-olds are made to run around if they complain: "If children feel cold, the adults activate them." During fall and spring, they have "tent weeks" when they sleep outside in the forest.

It was in one such day care that Lehtimäki did a study comparing the skin microbiomes in these kids to those in a more traditional day care and found, unsurprisingly, greater biodiversity. I was not imagining that kind of day care going over well in the U.S., where we coddle our youth and litigate when our kids die of hypothermia. But actually, nature-based programs are springing up. One can be found down the road from my home in Park Slope, Brooklyn: Brooklyn Forest, a parent-child class for preschoolers meant to help kids build "a meaningful connection to nature and wildlife . . . with vigorous physical work and nourishing food; with simple rhythms and constant singing; with feeling at home in the forest."

This connection happens in Prospect Park. Such goals are in keeping with the intent of the park when it was built 150 years ago—except for the singing, which was not explicitly stated in the vision of the designer, Frederick Law Olmsted. Known to some as the "father of landscape architecture," he made his name as the visionary behind Manhattan's 843-acre Central Park. In the only slightly smaller Prospect Park, in the center of Brooklyn, a series of signs hanging in a gazebo tells the history of the unlikely movement. In the mid-nineteenth century, a large public park was a new idea: "The poverty, social unrest, poor sanitation, and epidemics that plagued American cities convinced many city leaders that urban life was too stressful for its citizens," one of the signs reads. "Prospect Park was created to bring the healthy, calming effect of nature to all the citizens of Brooklyn."

The passive voice is doing a lot of work there; creating these places was a long and expensive undertaking. Though the cost of building Prospect Park was originally estimated at $300,000, the seven-year project ultimately came in at more than $5 million (more than $150 million today). Olmsted and his team meticulously mapped each acre on scrolls and designed them to perfectly capture an ideal "natural" landscape. Prospect Park, with its forests and meadows speckled with ornate tunnels and bridges over perfectly meandering creeks, unpretentious rose gardens and serendipitous waterfalls, was deliberately designed to look like it woke up like this.

The project relied on a vision that would reconceptualize what was missing from modern life, what would drive the epidemics of the future, and what could be done to stop them.

• • •

An idealistic millennial spirit living in the 1840s, Frederick Law Olmsted spent decades bouncing between professions, searching for something meaningful. Raised Puritan with enough family means to explore multiple paths without committing, he sailed to China as an apprentice seaman and then farmed on Staten Island, all the while looking for a way to contribute to the world despite the old problem that he, according to one biography, "found pursuing a career for money distasteful."

And so, he became a journalist. What followed was a *Forrest Gump*–type series of careers that put him at the center of the country's most consequential war, the design of many major cities, the very concept of what a city is, and the developing role of government in health. Olmsted's early journalism mostly involved newspaper stories about slavery in the Southern states, but it was a trip by foot through England in 1850 that awakened him to his calling. The country had just opened its first publicly funded park, called Birkenhead, in a suburb of Liverpool. He visited and had what his biographers describe as a sort of textbook epiphany. In the modern lingo of TED speakers, it was an "aha moment."

He noted the park's authentic combination of art and nature and the community that gathered within it. Olmsted was especially excited to discover that Birkenhead's beauty was shared "about equally by all classes," at a time when most parks tended to be located within private estates or, as in the case of Manhattan's Gramercy Park, locked behind gates.

To be a "melting pot" was ostensibly part of America's mission statement, but in practice, as lines were drawn between wealthy landowners and impoverished immigrants, there were fewer spaces for the melt. Years later, on his walks to work as an editor at *Putnam's Monthly*, Olmsted witnessed the birth of Lower Manhattan. What had been farmland a decade before was now a maze of hastily constructed low-rise buildings with multiple narrow, dark, extremely hot or freezing apartments. These would become known as tenements.

Though many have since been gut-renovated and sold for

millions of dollars, a few remain preserved as reminders of the health challenges that urban life made immediately clear. Touring New York's Tenement Museum, they seem almost spacious by New York standards—until the docent tells me that ten people would have been crammed into each apartment. The former residents of the one I toured were lucky to have three toilets in the backyard, which were shared with patrons of the pub on the first floor. Others left people to go in alleys and streets.

In the summer of 1857, Manhattan erupted in riots. In the four preceding decades, the population of the island had more than quadrupled. The space and resources that many had immigrated to pursue felt suddenly finite, and a sense of scarcity loomed. As the city had filled in the years leading up to the riots, city leadership had decided that the cure for social unrest was public space. It set aside a swath of the island for what would become the nation's first public park. Though it was to be built in an area where real estate values would one day near $2,000 per square foot—and where such square feet could be stacked on top of each other in high-rises—the city had the full support of its financial elites in setting aside *843 acres*, under the auspices of making New York a grand, enviable, global city.

In keeping with the competitive American spirit, the city held a contest for park designs. Olmsted partnered with architect Calvert Vaux (who, by many accounts, did most of the real work, but was a less public face). The pair was declared victorious in 1858 with something that went far beyond a park: a vision of artistic and cultural life, occupying an area so large that it was to include a Gothic castle in the center to keep wanderers oriented. Olmsted believed that huge public parks would be necessary to serve as the "lungs of the city," as the skies filled with industrial smog. His belief may have been based on the archaic miasma theory of disease (the one in which maladies like plague were believed spread by means of mysterious vapors), but the importance of clean air to health was also, of course, true. Miasma translates literally to *bad air* or *pollution.*

As technically inaccurate as miasma theory was, it also led to brilliant health innovations. Olmsted and Vaux emphasized well-drained land and waterways and "sanitary facilities"—or public bathrooms. In most of New York you have to buy a three-dollar shot of espresso just to use the bathroom. Restroom access is one of the reasons that some people I know maintain expensive gym memberships. Meanwhile, in Central Park there are twenty-one public restroom facilities.

This was part of a vision for social life that now feels surreal. Not only did the city build restrooms, it built elaborately tiled shrines to sanitation. The largest fountain in Central Park is actually an homage to the aqueduct that first brought fresh water to the city from upstate just sixteen years earlier. We once clearly knew the value of clean air and water, nature, and common spaces.

Central Park was estimated at over $500 billion in land value in 2005, a number that has surely continued to balloon with the city's real estate market. Though, of course, the value of the land would plummet if it were built upon, as would all other Manhattan property. The declines in health and community stand to be even greater.

Olmsted's work on Central Park brought him to the attention of Henry Bellows, a Unitarian minister from New York. At the onset of the Civil War, Bellows helped establish the U.S. Sanitary Commission to address the conditions of Union Army camps, and he recommended Olmsted to lead the new organization. The novel interdisciplinary team that Olmsted recruited would include not just physicians but an architect and engineer, theologians, philanthropists, and financial analysts.

Union generals were initially reluctant to have the Sanitary Commission redesign their camps, seeing the effort as a distraction. Periodic outbreaks of the extremely deadly smallpox virus, or of yellow fever, might get public attention. But other conditions—including tuberculosis, malaria, pneumonias, and diarrheal diseases—were seen as inevitable daily realities.

That changed in the wake of the Union Army's defeat at Bull Run in 1861. Lincoln grew desperate to turn the tide. Olmsted's Sanitary Commission argued that the soldiers' living conditions had contributed to the rout. In his report to the president, Olmsted wrote that troops were demoralized by fatigue, heat, and "want of food and drink." The army was not known for lifestyle amenities, of course, but the camps had grown particularly squalid. The generals and command in Washington had been concerned with little beyond arming the men and keeping them ambulatory. Anything more was either wasteful or frivolous—certainly not a matter of strategy.

Olmsted argued that it was. He pushed for investment in keeping people well—to keep them functioning properly. This was in a very rough sense arguing for one of history's first workplace wellness programs, like the treadmill desks at Google or the *Huffington Post*'s nap pods. To fight effectively, Olmsted urged prioritization of preventive medicine and soldiers' health.

When the U.S. government finally allowed the commission access to the camps, Olmsted and his colleagues demanded changes in location to minimize contamination of food and water—to ventilate spaces where soldiers were living in close quarters, and to allow food to be stored and prepared safely. With these changes came a surge in morale and performance. The lesson spread into later conflicts. Olmsted's Sanitary Commission became the core of the American Red Cross.

This would be just an early part of his total effect on public health—and his role in shaping the appearance and culture of the nation, bringing it together in its time of deepest division.

• • •

Across the ocean, around the same time, the British were fighting back Russian expansion into Crimea. Wounded and ill in the unfamiliar climate, the ranks were being decimated by infectious diseases. By some accounts, ten times more soldiers were dying of infectious diseases (typhus, typhoid, cholera, and dysentery) than were dying in battle.

London mustered a volunteer brigade of nurses led by Florence Nightingale. When the nurses arrived at the military hospital, they found soldiers wounded and dying amid horrifying conditions. Hospitals in the 1800s were where people went not to be healed, but to suffer and die. They were a sort of anteroom to hell—or heaven, sorry, heaven. Wherever your head is.

Nightingale found the places reprehensibly dank. One needed no understanding of germ theory to see men's beards and linens swarming with lice and fleas, errant excrement, and rats scattering. Nightingale believed the men needed air. The British government sent a novel "Sanitary Commission" for reinforcement, and she directed them to carve new doors and windows to get the breeze flowing through the rooms.

Almost instantly the condition of the men improved—though no one knew exactly why. London's *Times* referred to Nightingale as a "ministering angel." Though the men reportedly thought her work trivial in a time of war, as the death rate began to fall—by one report from 40 percent to 2 percent—the military leadership and even the queen took notice.

Nightingale became an advocate for the improvement of care and conditions in hospitals. The Crimean tale spread and changed the way many institutions operated. One of her books, *Notes on Hospitals*, makes the essential argument for better ventilation, more windows, drainage, and less-cramped conditions—in other words, solutions that predicted all the challenges of modern cities and hospitals.

Though Nightingale had a moment as possibly the world's first hygiene influencer, by the end of the century the roots of germ theory would take hold. Nightingale's advocacy of open air and exposure to nature got lost in the crusade to eliminate all microbes. As legitimate fears of contamination and infection became paramount, cleanliness became synonymous with sterility. Modern hospitals vied to provide apparently pristine conditions and personal privacy. People were put into small rooms with limited ventilation. Windows were small and kept closed in the interest of heating bills and a general idea of sterility that did not involve whatever the wind might blow in.

Only in recent years have the imperfections of this approach begun to be understood. Airflow-modeling techniques have tracked outbreaks within hospitals that could well have been prevented with simple open windows. And the understanding of the microbiome makes it apparent that the concept isn't just about letting pathogens escape, but about letting beneficial and inconsequential microbes in.

As the microbiologist Jack Gilbert put it at a 2012 conference:

"There's a good bacterial community living in hospitals, and if you try to wipe out that good bacterial community with sterilization agents and excessive antibiotics, you actually lay waste to this green field, this protective layer, and then these bad bacteria can just jump in and start causing hospital-borne [or hospital-mediated] infections."

This microbial world makes clear that health is a balance—balance between personal and public health; balance between being too exposed and too isolated. Among the wealthy, the tendency toward isolationism often wins out. When I interviewed the entrepreneur (and author of "more than 86 books") Deepak Chopra in 2017, he had just launched a new business selling "wellness real estate." The multimillion-dollar luxury apartments in New York and Miami have elaborate air filtration systems and countertops that are supposed to kill all microbes.

If this wellness real estate business were based in evidence, it would much more likely take the opposite approach. Apartments would maximize social bonding and exposure. Wellness housing might even offer to populate spaces and surfaces with benign and beneficial microbes. You can now buy bacterial room sprays (Goop now sells one, launched since I started writing this book) and "home-biotic" devices for home use that mist bacteria into the air. Better proven and more cost-effective, Gilbert says: open the windows.

That is, as long as air pollution levels make this possible.

• • •

After five dark months in tiny, overpriced apartments, the first warm weekend in Prospect Park feels like everyone is simultaneously and aggressively coming back to life. Though I live in a 250-

square-foot apartment, I have 526 acres that also feel like mine—but better, because wandering them alone would be no fun at all. And the cost of maintaining the boathouse would get on my nerves. The most popular attraction is the simple interior loop, a paved car-free road more than three miles around. The opportunity to run or bike uninterrupted by cars or road crossings is a rarity even in suburbs. It allows you to lose yourself in thought. I do it most days, and it works like running almost nowhere else in the city. Some of this book was written at picnic tables in the park; I took calls with researchers while walking the trails. I partook of the public restrooms.

The most urgent needs in the domain of human health today, globally, are clean air and water. Close behind are toilets, social connection, exposure to nature, and an active life in a safe environment. For a person who wants to maximize their impact on human health, any of these targets present opportunities that require no medical degree or attendant student loan debt. And all of these ideas—and the answer to how best to take care of our skin—come together in the park.

Olmsted's living work remains the backbone of New York City—he was among the designers of Union Square, Morningside, and Riverside parks. His vision is also carved into other cities across the country. As the walls of urbanization closed in, Olmsted traversed the country seeding it with parks. Certain that the country's rapid growth would compound in the cities, he protected spaces for the classes to mix and the air and water to circulate. He predicted that Central Park would one day lie at the heart of a metropolis and saw his work as preservation for future generations.

And so he gave us, among other things: the grounds of the U.S. Capitol and the National Zoo in DC; the grounds of the 1893

World's Fair in Chicago; the campus of Stanford University; as well as parks in Louisville, Atlanta, and Buffalo. He eventually moved to Boston, where he encircled the city in a corridor of green space totaling around 1,000 acres. A chain of nine parks stretches from Dorchester to Back Bay and Boston Common, seven miles long in total. Olmsted wanted to name it the Jeweled Girdle. He was ultimately, thankfully, steered toward the name Emerald Necklace. The parks are connected by what he called "pleasure roads," a concept that would become known as "parkways."

At the same time that these living embodiments of public health were being constructed—along with public systems of water and sanitation that would add decades to life spans around the world—private corporations also began developing medications to control and treat diseases. In combination, the medicine and public-health communities gained control over smallpox, polio, and diphtheria. In North America, malaria and yellow fever were virtually eradicated.

With the ability to treat and cure individuals with diseases, the future of medicine seemed bright. Doctors were no longer palliating, covering up pain and amputating limbs, but curing. They were treating disease processes at a cellular level. The focus of health investment shifted to individual treatments, and over time these grew more and more specific and expensive. We have now entered an era of "personalized medicine." In 2016, I moderated a panel kicking off the Precision Medicine Initiative, when President Obama and the nation's leading federal scientists announced a commitment to investing in treatments that would be tailored to every individual's specific biology.

I was only mildly skeptical at the time. Now I realize this way of thinking is turning our attention away from the much more urgently needed commitment to building the active, collaborative, engaged, social lifestyles that are the foundation of health. The two approaches are not mutually exclusive, of course, but we have swung too far toward self-care, dietary supplements, prescriptions, skin care, personal trainers, chiropractors, gurus, and medicines made to match our DNA. Soon we could be matching them to our microbiomes, too.

As I look now at the costs of this individualistic approach across so many multibillion-dollar industries—pharmaceuticals, cosmetics, supplements—I am unconvinced that more investment in treatments that work for an ever-smaller number of people should be a top priority. These approaches to health, by design, treat symptoms and diseases once they have occurred. They rarely work to prevent disease—the market incentive is to maximize use of a product, not to minimize it.

Today in the rapidly growing cities of the developing world, millions of people endure living conditions comparable to those of the tenements of the Lower East Side. The old diseases of deprivation are still rampant, but they are paired now with diseases of abundance. Parts of the world are in desperate need of basic sanitation and hygiene, food and water, while others have hoarded resources to their own detriment.

A century and a half after Olmsted's vision for public health led him to build parks, we've built fences and walls, and many of us live on cul-de-sacs with lawns full of pesticides and herbicides that are intended to kill everything but one particular species of grass. Our bathrooms are lined with bottles and creams and sprays that promise to protect us from the outside world, and now, increasingly, to restore the ecosystems we have washed away.

In 1950, 751 million people lived in cities. Today that number is 4.2 billion. By 2050, there is projected to be 2.5 billion more in those eroded, soaring cities. Each person will have less exposure to nature, to sunlight, to space for exercise. As we change our worlds, we change our bodies. The old duality between environmental health and human health is obsolete.

This is why I felt so absurd to be standing seven stories over Bryant Park, waiting to get hyaluronic acid and expensive serums rubbed into my face, alone behind a window that wouldn't open.

I'm not suggesting everyone should give up on skin care or quit showering. More than anything, this whole experiment helped me understand their value. These habits are profoundly personal, and it's important that decisions about them are made with maximal autonomy. This requires information, though, and this is where the landscape is skewed heavily toward systems that don't always work in our favor. This book is meant only to offer an alternative perspective on how our personal care habits affect our bodies and the communities on and around us. Advancing public health depends on constantly questioning the systems that presume to set the standards for what we consume and how we behave. It depends on understanding that we are all in this together, and that no challenges will be solved by sealing ourselves away from the exposures that sustain us, chasing some ineffable state of being clean.

EPILOGUE

One of the most dangerous places a person can be, in terms of serious infectious diseases, is a hospital.

It's possible the most contaminated thing in a hospital is the people who go from room to room touching everyone. Though there are now ordinances that require doctors to wash their hands, their white coats are in most cases rarely washed. People ask me why doctors wear scrubs in public, and how far it's best to stay away from these people. I can't give an exact distance. It's not an ideal practice, and it's definitely possible that these scrub-wearers are spreading pathogenic microbes into the community. But probably more pressing is the fact that doctors and other health care workers spread infections around hospitals. According to the CDC, every day one in thirty-one patients in an American hospital gets an infection from some exposure while there.

When I was in residency, I did a study at our small Cambridge hospital to try to understand how patients want their doctors to dress. I distributed a survey that had photos of me in various states of attire: scrubs with a white coat, scrubs alone, shirt and tie, no tie, with and without a white coat, etc. It turned out that people's preferences were all over the place. Some were more likely to trust a doctor in formal attire, even though they knew concerning things, like that ties don't get cleaned after every use. Others wanted a doctor in scrubs because they seem more ready and willing to do actual work. Scrubs are also washed more often than coats or ties.

Among the things I learned was that even though these adornments are somewhat dangerous, they also bring actual value to the patient-doctor interaction. Some people see them as status symbols that create barriers to communication and trust; others see them as signs of professionalism and confidence. These are elements that would be lost if infection-control officers at some hospitals had their way and asked doctors to wear disposable full-body suits and respirators into every room.

That sort of extreme protection would also make people (who suddenly find themselves referred to as "patients") in hospitals feel only more dehumanized than they often do. The basic requirement that doctors wash or sanitize their hands before and after touching any patient can already make people feel like some disgusting specimen. Sometimes such precautions are vital. But other times they serve to disconnect and alienate.

The psychological messages that we send to one another—as doctors and otherwise—are a reason to maintain basic standards of cleanliness. I still don't "shower" in a totally traditional sense, but I would never wear a white coat two days in a row without cleaning it. I would never wear a tie in a health care setting unless I washed it just as often as all my other clothes. Many mornings I turn on the water and lean my head in to get my hair wet, because otherwise it looks smashed and swirled in the way of permanent bed head, and I don't think people find this respectful.

Over the course of writing this book, I realized that vanity is a small part of explaining the ways we care for our skin. So is simply not offending others. In many ways, we clean and adorn ourselves as a way of honoring others. This is obvious in the act of, say, wearing a suit to a funeral but it plays out more subtly every day when we show we made an effort to be presentable, be it for a date or a meeting or just getting coffee. This was the main thing on my mind when I'd go out with bed head or smelling bad: that gnawing feeling, less of being judged than of seeming disrespectful to everyone who took the time to make themselves up for everyone else.

Considering the frequency and severity of hospital-acquired infections and in-hospital mortality related to medical errors, a lot of days I wondered how much good I was doing as a doctor. Whatever good is done by the health care industry comes at an annual cost of more than $3.5 trillion in the U.S. alone. The number is approaching 20 percent of our gross domestic product. In 2018, health care spending averaged $11,172 per person.

From pharmaceuticals to soaps and other personal care products, Americans are clearly overpaying for—and overusing—products and services that are supposed to make us healthier. The pattern of consumption is unsustainable, and much of it may be doing more harm than good. The greatest advances were those basic gestures at exposing people to nature—letting us have space to move, clean air to breathe, people to socialize and build relationships with, and plants, animals, and soil that bring us the microbes we evolved to be covered and sustained by.

Learning about the new understanding of the skin microbiome over the past few years, I was reassured that it's a pretty brilliant product of millions of years of evolution, a superorganism composed of trillions of other organisms that were doing fine before we came along and will do fine after we are gone. The ecosystem does not *need* to be maintained in any elaborate way that we didn't already know made our skin look good: sleeping and eating well, minimizing anxiety, and spending time in nature.

Even more reassuring for me has been the discovery that there are good health reasons to spend time in nature, to have pets, and to be social. Our instincts have been mostly right: we somehow know that going hiking is better than walking on a treadmill; that gardening is better than grocery shopping; that keeping house plants does something for us that makes it worth worrying about keeping them alive.

As clearly averse as I am to the idea of people being sold useless products based on false promises, I'm not without hope that things could get better. As I've learned about the history of soap marketing and what it did for germ theory and hygiene—popularizing notions that were otherwise tough sells—I've become sort of optimistic about what the skin probiotic concept will do. Skin probiotic products themselves may be a waste of money and time, and they may cause some bad reactions in people. But if it's in our nature to groom ourselves, and an inevitability that we will be sold products to do so, the overall narrative is moving in a healthier direction.

Clean may defy definition, but it is rife with meaning. It can imply isolation and sterilization, or plurality and diversity. Standards of acceptability are social, transient, and largely arbitrary. Considering our microbiomes, though, could shift more people toward a basic awareness of the fact that the ways we care for our skin never affect us alone. There are literal communities all over us and all around us. They affect everything we do, and everything we do affects them.

Ideally a quest to be clean involves worrying less about austere standards of sterility and instead embracing our complexities. The quest is to understand the world as an extension of ourselves. When we actively seek a balance of targeted hygiene and meaningful exposure to that world, the resulting sense of unity may be as close to the essence of clean as any definition I've yet found.

AFTERWORD

To be perfectly honest, I didn't expect that this book would come out in the heart of the worst pandemic in a century. It never so much as crossed my mind to begin to imagine that the book would be released in a moment of global panic about sterilizing surfaces that rendered Clorox wipes a precious commodity, and where procuring hand sanitizer required a *connection*.

I imagined that the year after the book was published might be spent discussing the merits and drawbacks of cleanliness, the nature of advertising in shaping beliefs about self and other, and deconstructing corporate indoctrination on the virtues of soap. Since I wrote the last sentences, millions of people have died of a disease that did not, at the time, even exist.

The pandemic has been an all-hands-on-deck moment for public health. Instead of being the author who people make fun of for not showering and spouting hippie nonsense about loving your microbes, I had to become the author who people made fun of for not showering but who also told them how to stay safe during a pandemic. I had to put on a tie for Zoom calls and go live on national TV from my couch. I found myself part of a chorus of doctors who would repeat, essentially: wash your hands, wear a mask, and stay away from other people.

Occasionally an interviewer would raise an eyebrow or two and say something to the effect of: "Aren't you the guy who said soap was bad? Which is it?" To which I wanted to respond with something about how no household product is wholly "good" or "bad," and that the nature of such labels depends on context, expectations of the consumer, and the society at large.

But instead I would stick to my talking points: "Wash your hands. Stay away from other people. Thank you for having me on your program."

It's impossible to imagine a more tragic coda to a book about microbial ecosystems. Despite all the technological advancements of modern medicine, a virus that is too small to be seen under a microscope—and that can be prevented by washing our hands, wearing cheap masks, and breathing fresh air—has bankrupted industries, overwhelmed health care systems, and brought that world to its knees. COVID-19 is caused by a strand of RNA encased in a rudimentary capsule; not enough even to constitute a living organism. Rather the individual viruses are referred to as "viral particles." These particles linger in the aerosols we emit when we breathe. We carry them on our skin and in the mucous membranes of our noses and airways, and they don't always harm us, but they can make us very dangerous to others. We can emit particles that land in the airways of others and send their immune systems into chaos.

This coronavirus (SARS-CoV-2) is clearly germane to all that this book is about, even despite it never appearing in the text. I believe the ideas in these pages, and the lessons learned while writing them, apply beyond any single disruption to the global infectious landscape. The pandemic has only put a lens to its importance.

Much has and will be written about the failures of the systems that led to and perpetuated the tragedy. I'd like to take a moment here to highlight a few lessons of the pandemic as they relate to this book. We have inordinate opportunity to learn from this collective experience, as we brace for the long-term consequences of the past year and a half.

As horrific as years like 2020 are, at least in brief moments there's something unifying about the clarity of crisis. Emergency situations demand that we shear away the (often trivial) uncertainties that so often consume us. Small anxieties that once kept us awake at night now feel petty and irrelevant. Health decisions like whether to put a probiotic into our armpits are contextualized, appropriately, as just not that important in the grander scheme of existence.

Instead of worrying about whether we're doing deodorant right or wrong, we can put our faith in a few straightforward directives. We know that if we follow them, we are safe, and have survived another day. Put on mask, stay outside, do not come close to others, perform essential mission (procuring beans or soap), return home, wash hands. Repeat as needed. We are invited to appreciate the fundamentals that day-to-day life would have us take for granted: if you and your loved ones have survived another day and are in good health, that's something to appreciate as fully as possible.

In the beginning, there was a cohort of people who said they found the structure of pandemic life to be cathartic. But whether it took days or weeks or months to grow to loathe it, the unsustainability was clear to everyone. Whatever brief comfort comes of the rigid structure fades, and the exact same directives create a sense of being constrained and overprotected. Where we once wanted clear directives, now we want to break those directives.

This pattern plays out in all sorts of health-related behaviors, from joining a gym to changing how we eat. It's in our nature to be drawn to rigid regimens that promise health and safety, and then to chafe against them and want to break free. Life is a sort of sine wave where we bounce between these poles.

With that in mind, emerging from more than a year of social distancing, this is a moment for hope. Readers have asked me whether this experience will forever warp our perception of cleanliness: whether people will never shake hands again, never cram ourselves into crowded restaurants, never ride in a subway car without a mask (or never ride in a subway car at all). I'm sure that for some people—especially those who suffered severe cases of COVID-19, or who lost family members—that memory will forever create unease in close quarters. Mass-casualty events like this create post-traumatic symptoms like those after combat. Many of us will be hyper attuned to any scenario in which we share microbes with others, and some may never feel fully comfortable in close proximity again.

For many others, though, we can expect that the sine wave will curve rapidly upward. We will appreciate parts of life that we took for granted before the pandemic, like the small interactions with people who aren't close friends or family—whom we barely know at all. Worries about the social dynamics of meeting new people in prepandemic life may be replaced by a basic thrill at the opportunity. I, for one, spent much of the winter of 2020 thinking about what I might say when I next saw my mail carrier or the people who work in my grocery store. I found myself looking forward to one day having a full conversation, without masks. I doubt any of them felt the same, but you get the idea.

For all the aspirational talk of "returning to normal," that's not going to happen. Nor should it. We should not strive to go backward, but to build on the lessons of this disaster to ensure that it never happens again. Our conceptualizations of ourselves and our vulnerabilities—of the limits of modern medicine, the inexorable link between science and politics, and the nature of truth itself—should all have undergone permanent shifts. This will leave us better prepared to identify and stop the next pandemic, and to address all the systemic issues on which it shined a spotlight. To the question of who could have predicted a pandemic of this magnitude, the answer is a resounding: everyone who studies infectious diseases. They knew this could happen at any time, and some day would. I was among them. I even mentioned in my 2016 book that virologists have told me this personally: there will be another pandemic of the magnitude of the 1918 influenza that killed millions. The only question was *when* this would happen. There is hopefully a way to fully enjoy and appreciate the social elements that make life great—without fear or constant anxiety—while also knowing that another pandemic of this magnitude of greater could very well happen at any time.

This doesn't mean we need to continue masking and distancing, or that we should cancel Coachella forever (at least, not for this reason). But it does mean we need to invest in public health, reliable media, and equitable systems of housing, nutrition, and criminal justice. We need to maintain strong global coalitions so that the world can respond immediately and effectively at the earliest signs of the next virus.

Among the many lessons of the pandemic is the reminder not to take our health for granted, or to rely on modern healthcare to keep us from getting sick. This has been a humbling moment for the medical profession, and particularly in the U.S.—where we spend so much more on health care than does any other country, and yet have incurred far greater losses of life—the fundamentals of health come down to our daily decisions and the creation of communities that keep us healthy. The system of hospitals and prescriptions and surgeries is only a safety net, and an unreliable one at that.

The pandemic does not change the basic fact that the microbiome is still more *us* than *them*, and embracing it continues to hold enormous promise. Because a virus is not technically alive, it is not a "normal" part of any biome. Some viruses are found on and in people commonly, and apparently do no harm, but nor do they seem essential to our functioning. If the coronavirus stays on your skin, it will degrade and vanish. It is a malicious invasive species, and our goal is to eradicate it from the world. There is no nuance to be found here.

But our efforts to prevent and stop its takeover could include more than just attempting to isolate and sterilize ourselves; they could include fortifying our biomes and immune systems in order to make them less prone to invasion. This approach is still hypothetical, and, for now, the path of isolation and sterilization is necessary for us all.

This leads to the final and probably the most common question I've heard from people who've read the book during the pandemic: How will the public-health measures that we deployed to stop the virus affect our immune systems in the longer term? It's a great question, and it does seem like something I should be able to answer. After all, I wrote about the value of proximity and touch, and of sharing microbes and chemical signals. All that remains true, even while, in the short term, it has been absolutely necessary to distance and sterilize ourselves even more than we once did.

The difficult fact is that, as in the cases where interviewers who pressed me to say whether soap was good or bad, the answer is that both of these things can be true. This is a tragedy from which no one comes out unscathed. Even if you didn't catch the virus, and didn't lose your job, and everyone you love remained relatively fine, you should emerge from this a different person. The extreme physical isolation that this pandemic has necessitated of so many of us will surely have some effects on our long-term health (on our immune systems and beyond). A complex network of ramifications that will include cardio-metabolic disease, substance use disorders, anxiety, depression, insomnia, and so much more. If our skin gets worse in the midst of all this, it will be difficult to disentangle any particular cause. Amid so much uncertainty, it has proven vital to stay grounded in the basics of what we know to constitute health. When we're able to focus on movement, purpose, connection, and sleep, everything else usually falls into place.

This doesn't mean that it was *bad* to shelter in place when necessary, and to distance ourselves from others. It was, in the context of possible options, the best choice. Our challenge now is to create a world where this is not the best choice. It's within our capacity. The pandemic has revealed very little that is new. It has only brought preexisting contrasts into sharper relief. It has provided a concrete framework for what must be done to stay safe. But unlike the narrow, rigid directives of midpandemic life, the list of what must be done is at once clear and expansive.

The disease, like so many others, disproportionately affected poor and minority communities. The policies that keep those communities poor only served to exacerbate the disparities in wealth and resources over the course of the pandemic. Much of the public messaging about COVID prevention centered on individual behaviors like distancing and masking—which have been good and important tools. But the real problems and solutions are systemic. They will require constant attention, even after the sense of immediate threat from this virus has lifted. We cannot assume that a pandemic of this scale will not happen again for another century. We must build a world that is ready for it to happen tomorrow.

The scope of the problem sometimes feels impossibly vast. I've felt a lot of despair. I've tried to remind myself to take things a day at a time. The only way out of despair is by feeling I've made a difference—however small. I can't make a difference if I'm paralyzed by the futility of it all. This has also, in moments, led me back to thinking small.

Of all things, during the pandemic, I have found solace in showers, most mornings.

I still don't use products (soap or shampoo) but standing under the water each day has come to be one of the few grounding rituals I consciously undertake. In a world without so many other context cues about when to work and when to relax, the act of briefly standing under the water has come to play a vital role in delineating night from day. I get dressed in "real" clothes, and I wear them until the end of the day. I run in Prospect Park. I make dinner. I try not to remain in work mode after that. I try to take time to keep perspective on the brevity of my opportunity to be alive as a part of this complex, sometimes horrible ecosystem. We all temporarily inhabit a planet of microbes, and we are guests here. Eventually they will have the planet to themselves again. They have been around before us and will persist long after us. The question is not how to fit them into our worlds, but how we can fit into theirs.

ACKNOWLEDGMENTS

This book is dedicated to my parents, Nancy and Jim, and to my grandmother Norma.

It exists primarily because of the brilliance and tireless work of my wife, Sarah Yager, and editor, Courtney Young.

It would also not have been possible without the generosity and wisdom of the many sources, colleagues, and microbe enthusiasts who shared with me their time, ideas, insight, research, and personal hygiene habits. I'm particularly indebted to the works of Luis and Fortuna Spitz, Val Curtis, Graham Rook, Jenni Lehtimäki, Julie Segre, Julia Scott, Jack Gilbert, Rob Dunn, Elizabeth Poynter, Katherine Ashenberg, and Justin Martin, and to the insight and time of Alicia Yoon, Autumn Henry, Emily Kreiger, Rachel Winard, Julia Oh, Annie Gottlieb, Jane Cavolina, Adina Grigore, David and Michael Bronner, Avi Gilbert, Eric Lupfer, Kelly Conaboy, Leah Finnegan, Mariam Gomaa, Jackie Shost, and Katie Martin, among many others. I am also grateful for the guidance and support during the writing of the book from Victoria Costales, Howard Forman, David Bradley, Sten Vermund, Adrienne LaFrance, Jeffrey Goldberg, Ross Andersen, and Paul Bisceglio. Also, to everyone else who shared physical space with me during the experimentation processes described in these pages.

SELECTED REFERENCES

1. IMMACULATE

Abuabara, Katrina, et al. "Prevalence of Atopic Eczema Among Patients Seen in Primary Care: Data from the Health Improvement Network." *Annals of Internal Medicine* 170, no. 5 (2019): 354–56. https://doi.org/10.7326/M18-2246.

Armelagos, George, et al. "Disease in Human Evolution: The Re-emergence of Infectious Disease in the Third Epidemiological Transition." *AnthroNotes* 18 (1996): 1–7. https://doi.org/10.5479/10088/22354.

"Base Price of Cigarettes in NYC Up to $13 a Pack." Spectrum News NY1, June 1, 2018. https://www.ny1.com/nyc/all-boroughs/health-and-medicine/2018/06/01/new-york-city-cigarettes-base-price.

Bharath, A. K., and R. J. Turner. "Impact of Climate Change on Skin Cancer." *Journal of the Royal Society of Medicine* 102, no. 6 (2009): 215–18.

Chauvin, Juan Pablo, et al. "What Is Different about Urbanization in Rich and Poor Countries? Cities in Brazil, China, India and the United States." *Journal of Urban Economics* 98 (2017): 17–49.

Chitrakorn, Kati. "Why International Beauty Brands Are Piling into South Korea." Business of Fashion, December 19, 2018.

Chung, Janice, and Eric L. Simpson. "The Socioeconomics of Atopic Dermatitis." *Annals of Allergy, Asthma and Immunology* 122 (2019): 360–66. https://www.ncbi.nlm.nih.gov/pubmed/30597208.

Clausen, Maja-Lisa, et al. "Association of Disease Severity with Skin Microbiome and Filaggrin Gene Mutations in Adult Atopic Dermatitis." *JAMA Dermatology* 154, no. 3 (2018): 293–300.

Dréno, B. "What Is New in the Pathophysiology of Acne, an Overview." *Journal of the European Academy of Dermatology and Venereology* 31, no. 55 (2017): 8–12. https://doi.org/10.1111/jdv.14374.

Garcia, Ahiza. "The Skincare Industry Is Booming, Fueled by Informed Consumers and Social Media." CNN, May 10, 2019.

Hajar, Tamar, and Eric L. Simpson. "The Rise in Atopic Dermatitis in Young Children: What Is the Explanation?" *JAMA Network Open* 1, no. 7 (2018): e184205.

Hamblin, James. "I Quit Showering, and Life Continued." *The Atlantic*, June 9, 2016. https://www.theatlantic.com/health/archive/2016/06/i-stopped-showering-and-life-continued/486314/.

Hou, Kathleen. "How I Used Korean Skin Care to Treat My Eczema." The Cut, August 15, 2019. https://www.thecut.com/2018/02/how-i-used-korean-skin-care-to-treat-my-eczema.html.

Kusari, Ayan, et al. "Recent Advances in Understanding and Preventing Peanut and Tree Nut Hypersensitivity." F1000 Research 7 (2018). https://doi.org/10.12688/f1000research.14450.1.

Laino, Charlene. "Eczema, Peanut Allergy May Be Linked." WebMD, March 1, 2010. https://www.webmd.com/skin-problems-and-treatments/eczema/news/20100301/eczema-peanut-allergy-may-be-linked#1.

Mooney, Chris. "Your Shower Is Wasting Huge Amounts of Energy and Water. Here's What You Can Do About It." *Washington Post*, March 4, 2015.

Nakatsuji, Teruaki, et al. "A Commensal Strain of *Staphylococcus epidermidis* Protects Against Skin Neoplasia." *Science Advances* 4, no. 2 (2018): eaao4502. http://advances.sciencemag.org/content/4/2/eaao4502.

Paller, Amy S., et al. "The Atopic March and Atopic Multimorbidity: Many Trajectories, Many Pathways." *Journal of Allergy and Clinical Immunology* 143, no. 1 (2019): 46–55.

Rocha, Marco A., and Ediléia Bagatin. "Adult-Onset Acne: Prevalence, Impact, and Management Challenges." *Clinical, Cosmetic and Investigational Dermatology* 11 (2018): 59–69. https://doi.org/10.2147/CCID.S137794.

"Scientists Identify Unique Subtype of Eczema Linked to Food Allergy." National Institutes of Health, U.S. Department of Health and Human Services, February 20, 2019.

Shute, Nancy. "Hey, You've Got Mites Living on Your Face. And I Do, Too." NPR, August 28, 2014.

Skotnicki, Sandy. *Beyond Soap: The Real Truth about What You Are Doing to Your Skin and How to Fix It for a Beautiful, Healthy Glow*. Toronto: Penguin Canada, 2018.

Spergel, Jonathan M., and Amy S. Paller. "Atopic Dermatitis and the Atopic March." *Journal of Allergy and Clinical Immunology* 112, no. 6 suppl. (2003): S118–27.

Talib, Warnidh H., and Suhair Saleh. "*Propionibacterium acnes* Augments Antitumor, Anti-Angiogenesis and Immunomodulatory Effects of Melatonin on Breast Cancer Implanted in Mice." *PLoS ONE* 10, no. 4 (2015): e0124384.

Thiagarajan, Kamala. "As Delhi Chokes on Smog, India's Health Minister Advises: Eat More Carrots." NPR, November 8, 2019.

2. PURIFY

Ashenburg, Katherine. *The Dirt on Clean: An Unsanitized History*. Toronto: Knopf Canada, 2007.

Behringer, Donald C., et al. "Avoidance of Disease by Social Lobsters." *Nature* 441 (2006): 421.

Black Death, The. Translated and edited by Rosemary Horrox. Manchester Medieval Sources series. Manchester, UK: Manchester University Press, 1994.

Blackman, Aylward M. "Some Notes on the Ancient Egyptian Practice of Washing the Dead." *The Journal of Egyptian Archaeology* 5, no. 2 (1918): 117–24.

Boccaccio, Giovanni. *The Decameron*. Translated by David Wallace. Landmarks of World Literature. Cambridge, UK: Cambridge University Press, 1991.

Curtis, Valerie A. "Dirt, Disgust and Disease: A Natural History of Hygiene." *Journal of Epidemiology and Community Health* 61, no. 8 (2007): 660–64. https://doi.org/10.1136/jech.2007.062380.

———. "Hygiene." In *Berkshire Encyclopedia of World History*, 2nd ed., edited by William H. McNeill et al., 1283–87. Great Barrington, MA: Berkshire, 2010.

———. "Infection-Avoidance Behaviour in Humans and Other Animals." *Trends in Immunology* 35, no. 10 (2014): 457–64. http://dx.doi.org/10.1016/j.it.2014.08.006.

———. "Why Disgust Matters." *Philosophical Transactions of the Royal Society B* 366, no. 1583 (2011): 3478–90. https:doi.org/10.1098/rstb.2011.0165.

Fagan, Garrett. *Bathing in Public in the Roman World*. Ann Arbor: University of Michigan Press, 2002.

———. "Three Studies in Roman Public Bathing." PhD diss., McMaster University, 1993.

Foster, Tom. "The Undiluted Genius of Dr. Bronner's." *Inc.*, April 3, 2012.

Galka, Max. "From Jericho to Tokyo: The World's Largest Cities Through History—Mapped." *The Guardian*, December 6, 2016.

Goffart, Walter. *Barbarian Tides: The Migration Age and the Later Roman Empire*. Philadelphia: University of Pennsylvania Press, 2006.

Hennessy, Val. "Washing Our Dirty History in Public." *Daily Mail*, April 1, 2008. https://www.dailymail.co.uk/home/books/article-548111/Washing-dirty-history-public.html.

Jackson, Peter. "Marco Polo and His 'Travels.'" *Bulletin of the School of Oriental and African Studies* (University of London) 61, no. 1 (1998): 82–101.

Konrad, Matthias, et al. "Social Transfer of Pathogenic Fungus Promotes Active Immunisation in Ant Colonies." *PLoS Biology* 10, no. 4 (2012): e1001300.

Morrison, Toni. "The Art of Fiction," no. 134. Interview by Elissa Schappell and Claudia Brodsky Lacour. *Paris Review* 128 (Fall 1993). https://www.theparisreview.org/interviews/1888/toni-morrison-the-art-of-fiction-no-134-toni-morrison.

Poynter, Elizabeth. *Bedbugs and Chamberpots: A History of Human Hygiene*. CreateSpace, 2015.

Prum, Richard O. *The Evolution of Beauty: How Darwin's Forgotten Theory of Mate Choice Shapes the Animal World—and Us*. New York: Doubleday, 2017.

Roesdahl, Else, et al., eds. *The Vikings in England and in Their Danish Homeland*. London: The Anglo-Danish Viking Project, 1981.

Schafer, Edward H. "The Development of Bathing Customs in Ancient and Medieval China and the History of the Floriate Clear Palace." *Journal of the American Oriental Society* 76, no. 2 (1956): 57–82.

Schwartz, David A., ed. *Maternal Death and Pregnancy-Related Morbidity Among Indigenous Women of Mexico and Central America: An Anthropological, Epidemiological, and Biomedical Approach*. Cham, Switzerland: Springer International, 2018.

Yegül, Fikret. *Bathing in the Roman World*. New York: Cambridge University Press, 2010.

3. LATHER

Bollyky, Thomas J. *Plagues and the Paradox of Progress: Why the World Is Getting Healthier in Worrisome Ways*. Cambridge, MA: The MIT Press, 2018.

Cox, Jim. *Historical Dictionary of American Radio Soap Operas*. Lanham, MD: Scarecrow Press / Rowman and Littlefield, 2005.

"Donkey Milk." *World Heritage Encyclopedia*.

"Dr. Bronner's 2019 All-One! Report." https://www.drbronner.com/allone-reports/A1R-2019/all-one-report 2019.html.

Evans, Janet. *Soap Making Reloaded: How to Make a Soap from Scratch Quickly and Safely: A Simple Guide for Beginners and Beyond*. Newark, DE: Speedy Publishing, 2013.

Gladstone, W. E. *The Financial Statements of 1853 and 1860, to 1865*. London: John Murray, 1865.

Heyward, Anna. "David Bronner, Cannabis Activist of the Year." *The New Yorker*, February 29, 2016.

McNeill, William H. *Plagues and Peoples*. New York: Doubleday, 1977.

Mintel Press Office. "Slippery Slope for Bar Soap As Sales Decline 2% since 2014 in Favor of More Premium Options." Mintel, August 22, 2016.

"Palm Oil: Global Brands Profiting from Child and Forced Labour." Amnesty International, November 30, 2016. https://www.amnesty.org/en/latest/news/2016/11/palm-oil-global-brands-profiting-from-child-and-forced-labour/.

Port Sunlight Village Trust. "About Port Sunlight: History and Heritage."

Prigge, Matthew. "The Story Behind This Bar of Palmolive Soap." *Milwaukee Magazine*, January 25, 2018.

Savage, Woodson J. III. *Streetcar Advertising in America*. Stroud, Gloucestershire, UK: Fonthill Media, 2016.

"Soap Ingredients." Handcrafted Soap & Cosmetic Guild.

Spitz, Luis. *SODEOPEC: Soaps, Detergents, Oleochemicals, and Personal Care Products*. Champaign, IL: AOCS Press, 2004.

Spitz, Luis, ed. *Soap Manufacturing Technology*. Urbana, IL: AOCS Press, 2009.

Spitz, Luis, and Fortuna Spitz. *The Evolution of Clean: A Visual Journey Through the History of Soaps and Detergents*. Washington, DC: Soap and Detergent Association, 2006.

"Who Invented Body Odor?" Roy Rosenzweig Center for History and New Media. https://rrchnm.org/sidelights/who-invented-body-odor/.

Willingham, A. J. "Why Don't Young People Like Bar Soap? They Think It's Gross, Apparently." CNN, August 29, 2016.

Wisetkomolmat, Jiratchaya, et al. "Detergent Plants of Northern Thailand: Potential Sources of Natural Saponins." *Resources* 8, no. 1 (2019). https://doi.org/10.3390/resources8010010.

Zax, David. "Is Dr. Bronner's All-Natural Soap A $50 Million Company or an Activist Platform? Yes." *Fast Company*, May 2, 2013.

4. GLOW

Baumann, Leslie. *Cosmeceuticals and Cosmetic Ingredients*. New York: McGraw-Hill Education/Medical, 2015.

"Clean Beauty—and Why It's Important." *Goop*.

"Emily Weiss." The Atlantic Festival, YouTube, October 8, 2018.

Fine, Jenny B. "50 Beauty Execs Under 40 Driving Innovation." *Women's Wear Daily*, June 24, 2016.

Jones, Geoffrey. *Beauty Imagined: A History of the Global Beauty Industry* New York: Oxford University Press, 2010.

Strzepa, Anna, et al. "Antibiotics and Autoimmune and Allergy Diseases: Causative Factor or Treatment?" *International Immunopharmacology* 65 (2018): 328–41.

Surber, Christian, et al. "The Acid Mantle: A Myth or an Essential Part of Skin Health?" *Current Problems in Dermatology* 54 (2018): 1–10.

Varagur, Krithika. "The Skincare Con." The Outline, January 30, 2018.

Warfield, Nia. "Men Are a Multibillion Dollar Growth Opportunity for the Beauty Industry." CNBC, May 20, 2019.

Wischhover, Cheryl. "Glossier, the Most-Hyped Makeup Company on the Planet, Explained." Vox, March 4, 2019. https://www.vox.com/the-goods/2019/3/4/18249886/glossier-play-emily-weiss-makeup.

———. "The Glossier Machine Kicks into Action to Sell Its New Product." Racked, March 4, 2018. https://www.racked.com/2018/3/4/17079048/glossier-oscars.

5. DETOXIFY

Burisch, Johan, et al. "East–West Gradient in the Incidence of Inflammatory Bowel Disease in Europe: The ECCO-EpiCom Inception Cohort." *Gut* 63 (2014): 588–97. http://dx.doi.org/10.1136/gutjnl-2013-304636.

Dunn, Robert R. "The Evolution of Human Skin and the Thousands of Species It Sustains, with Ten Hypothesis of Relevance to Doctors." In *Personalized, Evolutionary, and Ecological Dermatology*, edited by Robert A. Norman (Switzerland: Springer International Publishing, 2016).

"FDA Authority Over Cosmetics: How Cosmetics Are Not FDA-Approved, but Are FDA-Regulated." U.S. Food and Drug Administration.

Feinstein, Dianne, and Susan Collins. "The Personal Care Products Safety Act." *JAMA Internal Medicine* 178, no. 5 (2018): 201–2.

"Fourth National Report on Human Exposure to Environmental Chemicals." U.S. Department of Health and Human Services Centers for Disease Control and Prevention, 2009.

Graham, Jefferson. "Retailer Claire's Pulls Makeup from Its Shelves over Asbestos Concerns." *USA Today*, December 27, 2017.

"Is It a Cosmetic, a Drug, or Both? (Or Is It Soap?)" U.S. Food and Drug Administration.

"More Health Problems Reported with Hair and Skin Care Products," KCUR, June 26, 2017.

Patterson, Christopher, et al. "Trends and Cyclical Variation in the Incidence of Childhood Type 1 Diabetes in 26 European Centres in the 25-Year Period 1989–2013: A Multicentre Prospective Registration Study." *Diabetologia* 62 (2019): 408–17. https://doi.org/10.1007/s00125-018-4763-3.

———. "Worldwide Estimates of Incidence, Prevalence and Mortality of Type 1 Diabetes in Children and Adolescents: Results from the International Diabetes Federation Diabetes Atlas, 9th edition." *Diabetes Research and Clinical Practice* 157 (2019). https://doi.org/10.1016/j.diabres.2019.107842.

Prescott, Susan, et al. "A Global Survey of Changing Patterns of Food Allergy Burden in Children." *World Allergy Organization Journal* 6 (2013): 1–12. https://doi.org/10.1186/1939-4551-6-21.

Pycke, Benny, et al. "Human Fetal Exposure to Triclosan and Triclocarban in an Urban Population from Brooklyn, New York." *Environmental Science & Technology* 48, no. 15 (2014): 8831–38. https://doi.org/10.1021/es501100w.

Scudellari, Megan. "News Feature: Cleaning Up the Hygiene Hypothesis." *Proceedings of the National Academy of Sciences of the United States of America* 114, no. 7 (2017): 1433–36. https://doi.org/10.1073/pnas.1700688114.

Silverberg, Jonathan I. "Public Health Burden and Epidemiology of Atopic Dermatitis." *Dermatologic Clinics* 35, no. 3 (2017): 283–89. https://doi.org/10.1016/j.det.2017.02.002.

"Statement on FDA Investigation of WEN by Chaz Dean Cleansing Conditioners." U.S. Food and Drug Administration, November 15, 2017.

Strachan, David. "Hay Fever, Hygiene, and Household Size." *British Medical Journal* 299 (1989): 1259–60. https://doi.org/10.1136/bmj.299.6710.1259.

Vatanen, Tommi. "Variation in Microbiome LPS Immunogenicity Contributes to Autoimmunity in Humans." *Cell* 165, no. 4 (2016): 842–53. https://doi.org/10.1016/j.cell.2016.04.007.

"Walmart Recalls Camp Axes Due to Injury Hazard." United States Consumer Product Safety Commission, October 3, 2018.

6. MINIMIZE

"Bacteria Therapy for Eczema Shows Promise in NIH Study." National Institutes of Health. U.S. Department of Health and Human Services, May 3, 2018. https://www.nih.gov/news-events/news-releases/bacteria-therapy-eczema-shows-promise-nih-study.

Bennett, James. "Hexachlorophene." *Cosmetics and Skin*, October 3, 2019.

Bloomfield, Sally F. "Time to Abandon the Hygiene Hypothesis: New Perspectives on Allergic Disease, the Human Microbiome, Infectious Disease Prevention and the Role of Targeted Hygiene." *Perspectives in Public Health* 136, no. 4 (2016): 213–24.

Böbel, Till S., et al. "Less Immune Activation Following Social Stress in Rural vs. Urban Participants Raised with Regular or No Animal Contact, Respectively." *Proceedings of the National Academy of Sciences* 115, no. 20 (2018): 5259–64.

Callard, Robin E., and John I. Harper. "The Skin Barrier, Atopic Dermatitis and Allergy: A Role for Langerhans Cells?" *Trends in Immunology* 28, no. 7 (2007): 294–98.

"Gaspare Aselli (1581–1626). The Lacteals." *JAMA* 209, no. 5 (1969): 767. https://doi.org/10.1001/jama.1969.03160180113016.

Gilbert, Jack, and Rob Knight. *Dirt Is Good: The Advantage of Germs for Your Child's Developing Immune System*. New York: St. Martin's Press, 2017.

Hamblin, James. "The Ingredient to Avoid in Soap." *The Atlantic*, November 17, 2014.

Holbreich, Mark, et al. "Amish Children Living in Northern Indiana Have a Very Low Prevalence of Allergic Sensitization." *Journal of Allergy and Clinical Immunology* 129, no. 6 (2012): 1671–73.

Lee, Hye-Rim, et al. "Progression of Breast Cancer Cells Was Enhanced by Endocrine-Disrupting Chemicals, Triclosan and Octylphenol, via an Estrogen Receptor-Dependent Signaling Pathway in Cellular and Mouse Xenograft Models." *Chemical Research in Toxicology* 27, no. 5 (2014): 834–42.

MacIsaac, Julia K., et al. "Health Care Worker Exposures to the Antibacterial Agent Triclosan." *Journal of Occupational and Environmental Medicine* 56, no. 8 (2014): 834–39. https://doi.org/10.1097/jom.0000000000000183.

Rook, Graham, et al. "Evolution, Human-Microbe Interactions, and Life History Plasticity." *The Lancet* 390, no. 10093 (2017): 521–30. https://doi.org/10.1016/S0140-6736(17)30566-4.

Scudellari, Megan. "News Feature: Cleaning Up the Hygiene Hypothesis." *Proceedings of the National Academy of Sciences* 114, no. 7 (2017): 1433–36.

Shields, J. W. "Lymph, Lymphomania, Lymphotrophy, and HIV Lymphocytopathy: An Historical Perspective." *Lymphology* 27, no. 1 (1994): 21–40.

Stacy, Shaina L., et al. "Patterns, Variability, and Predictors of Urinary Triclosan Concentrations During Pregnancy and Childhood." *Environmental Science and Technology* 51, no. 11 (2017): 6404–13.

Stein, Michelle M., et al. "Innate Immunity and Asthma Risk in Amish and Hutterite Farm Children." *New England Journal of Medicine* 375, no. 5 (2016): 411–21.

Velasquez-Manoff, Moises. *An Epidemic of Absence: A New Way of Understanding Allergies and Autoimmune Diseases*. New York: Scribner, 2012.

Von Hertzen, Leena C., et al. "Scientific Rationale for the Finnish Allergy Programme 2008–2018: Emphasis on Prevention and Endorsing Tolerance." *Allergy* 64, no. 5 (2009): 678–701.

Von Mutius, Erika. "Asthma and Allergies in Rural Areas of Europe." *Proceedings of the American Thoracic Society* 4 (2007): 212–16.

Warfield, Nia. "Men Are a Multibillion Dollar Growth Opportunity for the Beauty Industry." CNBC, May 20, 2019. https://www.cnbc.com/2019/05/17/men-are-a-multibillion-dollar-growth-opportunity-for-the-beauty-industry.html/.

7. VOLATILE

Baldwin, Ian T., and Jack C. Schultz. "Rapid Changes in Tree Leaf Chemistry Induced by Damage: Evidence for Communication Between Plants." *Science* 221, no. 4607 (1983): 277–79. https://science.sciencemag.org/content/221/4607/277.

Costello, Benjamin Paul de Lacy, et al. "A Review of the Volatiles from the Healthy Human Body." *Journal of Breath Research* 8, no. 1 (2014): 014001.

Emslie, Karen. "To Stop Mosquito Bites, Silence Your Skin's Bacteria." *Smithsonian*, June 30, 2015. https://www.smithsonianmag.com/science-nature/stop-mosquito-bites-silence-your-skins-bacteria-180955772/.

Gols, Richard, et al. "Smelling the Wood from the Trees: Non-Linear Parasitoid Responses to Volatile Attractants Produced by Wild and Cultivated Cabbage." *Journal of Chemical Ecology* 37 (2011): 795.

Guest, Claire. *Daisy's Gift: The Remarkable Cancer-Detecting Dog Who Saved My Life*. London: Virgin Books, 2016.

Hamblin, James. "Emotions Seem to Be Detectable in Air." *The Atlantic*, May 23, 2016.

Maiti, Kiran Sankar, et al. "Human Beings as Islands of Stability: Monitoring Body States Using Breath Profiles." *Scientific Reports* 9 (2019): 16167.

Pearce, Richard F., et al. "Bumblebees Can Discriminate Between Scent-Marks Deposited by Conspecifics." *Scientific Reports* 7 (2017): 43872. https://doi.org/10.1038/srep43872.

Rodríguez-Esquivel, Miriam, et al. "Volatolome of the Female Genitourinary Area: Toward the Metabolome of Cervical Cancer." *Archives of Medical Research* 49, no. 1 (2018): 27–35.

Verhulst, Niels O., et al. "Composition of Human Skin Microbiota Affects Attractiveness to Malaria Mosquitoes." *PLoS ONE* 6, no. 12 (2011): e28991. https://doi.org/10.1371/journal.pone.0028991.

8. PROBIOTIC

Benn, Christine Stabell, et al. "Maternal Vaginal Microflora During Pregnancy and the Risk of Asthma Hospitalization and Use of Antiasthma Medication in Early Childhood." *Allergy and Clinical Immunology* 110, no. 1 (2002): 72–77.

Capone, Kimberly A., et al. "Diversity of the Human Skin Microbiome Early in Life." *Journal of Investigative Dermatology* 131, no. 10 (2011): 2026–32.

Castanys-Muñoz, Esther, et al. "Building a Beneficial Microbiome from Birth." *Advances in Nutrition* 7, no. 2 (2016): 323–30.

Clausen, Maja-Lisa, et al. "Association of Disease Severity with Skin Microbiome and Filaggrin Gene Mutations in Adult Atopic Dermatitis." *JAMA Dermatology* 154, no. 3 (2018): 293–300.

Council, Sarah E., et al. "Diversity and Evolution of the Primate Skin Microbiome." *Proceedings of the Royal Society B* 283, no. 1822 (2016): 20152586. https://doi.org/10.1098/rspb.2015.2586.

Dahl, Mark V. "*Staphylococcus aureus* and Atopic Dermatitis." *Archives of Dermatology* 119, no. 10 (1983): 840–46.

Dotterud, Lars Kåre, et al. "The Effect of UVB Radiation on Skin Microbiota in Patients with Atopic Dermatitis and Healthy Controls." *International Journal of Circumpolar Health* 67, no. 2-3 (2008): 254–60.

Flandroy, Lucette, et al. "The Impact of Human Activities and Lifestyles on the Interlinked Microbiota and Health of Humans and of Ecosystems." *Science of the Total Environment* 627 (2018): 1018–38.

Fyhrquist, Nanna, et al. "*Acinetobacter* Species in the Skin Microbiota Protect Against Allergic Sensitization and Inflammation." *Journal of Allergy and Clinical Immunology* 134, no. 6 (2014): 1301–9.e11.

Grice, Elizabeth A., and Julie A. Segre. "The Skin Microbiome." *National Reviews in Microbiology* 9, no. 4 (2011): 244–53.

Grice, Elazbeth A., et al. "Topographical and Temporal Diversity of the Human Skin Microbiome." *Science* 324, no. 5931 (2009): 1190–92.

Hakanen, Emma, et al. "Urban Environment Predisposes Dogs and Their Owners to Allergic Symptoms." *Scientific Reports* 8 (2018): 1585.

Jackson, Kelly M., and Andrea M. Nazar. "Breastfeeding, the Immune Response, and Long-term Health." *Journal of the American Osteopathic Association* 106, no. 4 (2006): 203–7.

Karkman, Antti, et al. "The Ecology of Human Microbiota: Dynamics and Diversity in Health and Disease." *Annals of the New York Academy of Sciences* 1399, no. 1 (2017): 78–92.

Kim, Jooho P., et al. "Persistence of Atopic Dermatitis (AD): A Systematic Review and Meta-Analysis." *Journal of the American Academy of Dermatology* 75, no. 4 (2016): 681–87. https://doi.org/10.1016/j.jaad.2016.05.028.

Lehtimäki, Jenni, et al. "Patterns in the Skin Microbiota Differ in Children and Teenagers Between Rural and Urban Environments." *Scientific Reports* 7 (2017): 45651.

Levy, Barry S., et al., eds. *Occupational and Environmental Health: Recognizing and Preventing Disease and Injury.* 6th ed. New York: Oxford University Press, 2011.

Mueller, Noel T., et al. "The Infant Microbiome Development: Mom Matters." *Trends in Molecular Medicine* 21, no. 2 (2015): 109–17.

Myles, Ian A., et al. "First-in-Human Topical Microbiome Transplantation with *Roseomonas mucosa* for Atopic Dermatitis." *JCI Insight* 3, no. 9 (2018). https://doi.org/10.1172/jci.insight.120608.

Picco, Federica, et al. "A Prospective Study on Canine Atopic Dermatitis and Food-Induced Allergic Dermatitis in Switzerland." *Veterinary Dermatology* 19, no. 3 (2008): 150–55.

Richtel, Matt, and Andrew Jacobs. "A Mysterious Infection, Spanning the Globe in a Climate of Secrecy." *New York Times*, April 6, 2019.

Ross, Ashley A., et al. "Comprehensive Skin Microbiome Analysis Reveals the Uniqueness of Human Skin and Evidence for Phylosymbiosis within the Class Mammalia." *Proceedings of the National Academy of Sciences* 115, no. 25 (2018): E5786–95.

Scharschmidt, Tiffany C. "*S. aureus* Induces IL-36 to Start the Itch." *Science Translational Medicine* 9, no. 418 (2017): eaar2445.

Scott, Julia. "My No-Soap, No Shampoo, Bacteria-Rich Hygiene Experiment." *New York Times*, May 22, 2014.

Textbook of Military Medicine. Washington, DC: Office of the Surgeon General at TMM Publications, 1994.

Van Nood, Els, et al. "Duodenal Infusion of Donor Feces for Recurrent Clostridium Difficile." *New England Journal of Medicine* 368 (2013): 407–415. https://doi.org/10.1056/NEJMoa1205037.

Wattanakrai, Penpun and James S. Taylor. "Occupational and Environmental Acne." In *Kanerva's Occupational Dermatology*, edited by Thomas Rustemeyer et al. (Berlin: Springer, 2012).

Winter, Caroline. "Germ-Killing Brands Now Want to Sell You Germs." *Bloomberg Businessweek*, April 22, 2019.

9. REFRESH

Beveridge, Charles E. "Frederick Law Olmsted Sr." National Association for Olmsted Parks.

Borchgrevink, Carl P., et al. "Handwashing Practices in a College Town Environment." *Journal of Environmental Health*,

April 2013. https://msutoday.msu.edu/_/pdf/assets/2013/hand-washing-study.pdf.

Fee, Elizabeth, and Mary E. Garofalo. "Florence Nightingale and the Crimean War." *American Journal of Public Health* 100, no. 9 (2010): 1591. https://doi.org/10.2105/AJPH.2009.188607.

Fisher, Thomas. "Frederick Law Olmsted and the Campaign for Public Health." *Places*, November 2010.

Koivisto, Aino. "Finnish Children Spend the Entire Day Outside." Turku.fi, November 16, 2017. http://www.turku.fi/en/news/2017-11-16_finnish-children-spend-entire-day-outside.

Martin, Justin. *Genius of Place: The Life of Frederick Law Olmsted*. Boston: Da Capo Press, 2011.

National Archives. "Florence Nightingale." https://www.nationalarchives.gov.uk/education/resources/florence-nightingale/.

Olmsted, Frederick Law, and Jane Turner Censer. *The Papers of Frederick Law Olmsted, Volume IV: Defending the Union: The Civil War and the U.S. Sanitary Commission 1861–1863*. Baltimore: Johns Hopkins University Press, 1986.

Rich, Nathaniel. "When Parks Were Radical." *The Atlantic*, September 2016.

Ruokolainen, Lasse, et al. "Green Areas Around Homes Reduce Atopic Sensitization in Children." *Allergy* 70, no. 2 (2015): 195–202.

"Sanitation." UNICEF, June 2019. https://data.unicef.org/topic/water-and-sanitation/sanitation/.

"Sanitation." World Health Organization, June 14, 2019. https://www.who.int/news-room/fact-sheets/detail/sanitation.

"Trachoma." World Health Organization, June 27, 2019. https://www.who.int/news-room/fact-sheets/detail/trachoma. https://eportfolios.macaulay.cuny.edu/munshisouth10/group-projects/prospect-park/history/.

"WASH Situation in Mozambique." UNICEF. https://www.unicef.org/mozambique/en/water-sanitation-and-hygiene-wash.

INDEX